P9-DDC-974

ary

EARTHQUAKE PREDICTION

EARTHQUAKE PREDICTION

DAWN OF THE NEW SEISMOLOGY

BY DAVID NABHAN

Skyhorse Publishing

Disclaimer: The publisher is grateful for the use of photographs in this work reprinted from the archives of the United States Geological Survey (USGS), NASA, NOAA, FEMA, the US Army and the Library of Congress. Nothing is implied by their use, however, to represent in any way that those agencies and institutions endorse or recommend any part of this work.

Skyhorse Publishing books may be purchased in bulk at special discounts for sales promotion, corporate gifts, fund-raising, or educational purposes. Special editions can also be created to specifications. For details, contact the Special Sales Department, Skyhorse Publishing, 307 West 36th Street, 11th Floor, New York, NY 10018 or info@skyhorsepublishing.com.

Skyhorse® and Skyhorse Publishing® are registered trademarks of Skyhorse Publishing, Inc.®, a Delaware corporation.

Visit our website at www.skyhorsepublishing.com.

10 9 8 7 6 5 4 3 2 1

Library of Congress Cataloging-in-Publication Data is available on file.

Cover design by Rain Saukas
Cover photo credit: iStock

Print ISBN: 978-1-5107-2097-8
Ebook ISBN: 978-1-5107-2098-5

Printed in the United States of America

"Nabhan's theory deserves to be taken note of."

—Dr. Kate Hutton, chief seismologist, California Institute of Technology, to *Los Angeles Weekly,* April, 1997

"This is simple but brilliant observation, followed by reasonable applications; impossible to dismiss as coincidence."

—Dr. Cort Stoskopf, *Popular Science Radio,* 2013

"Knowing when an earthquake might strike may be as simple as tracking in the sky where the sun and moon are. If you believe that predicting earthquakes is impossible then everything you think you know about it could be wrong."

—Thom Hartmann, RT Television, May 20, 2014

"What if I were to tell you that nearly every single deadly Southern Californian earthquake in the past happened at dawn or at dusk and during a new or full moon?"

—Paul Magers, news anchor, CBS 2,
Los Angeles, April 30, 2014

"If proven this would be the 'Holy Grail' of seismology."

—Conan Nolan, news anchor, KNBC News, Los Angeles, 1997

"Nabhan's forecasts nailed three of the largest seismic events on the West Coast in 1996."

—Erin Aubrey, staff writer, *Los Angeles Weekly,* 1997

"There you go; seems to me you're *right*."

—George Noory, *Coast to Coast AM*, November, 2011

"David Nabhan is the man who says he's worked out how to predict earthquakes in Southern California . . . quakes happen during either a new or a full moon, and within three hours of dawn or dusk."

—*London Daily Mail*, June 1, 2014

"This is the first guy to go back and check what all those seismic events have in common? The idea that it just took somebody to say 'let's look and see if there's a pattern' is pretty shocking to me."

—Scott Cox, KERN TV/Radio June 3, 2014

"While some studies indicate that tidal effect may have an effect on smaller quakes, there's no evidence they have an effect on bigger quakes."

—Dr. Tom Jordan, director, Southern California Earthquake Center, to *AOL News*, 2010

"We've got a guy coming on who predicted a quake the last time on the show; I don't know what to make of this earthquake prediction stuff."

—Howard Stern, *Howard Stern Show*, January, 1998

"Ladies and gentlemen, you read it here first: Nabhan's near dead-on calculation for the August 20 shaker near Wrightwood."

—Theresa Moreau, *Pasadena Weekly*, 1998

"Whether or not Nabhan can predict earthquakes, he certainly predicts seismic skepticism."

—David Moye, *America Online News*, 2010

"It's such a no-no. Seismologists won't even talk about it!"

—Kate Delaney, *America Tonight*, 2010

"Nabhan is confident enough in his work to urge government leaders to take action and prepare for major tremors during certain periods of time."

—Andrew Ireland, *World Net Daily*, July 17, 2014

"There's a huge difference between a prediction and an advisory. But the hardships unprepared Valley residents face if the Big One hits? That's a promise."

—Jeff Jardin, *Modesto Bee*, June 13, 2013

"It's simply mind-boggling that with the evidence Nabhan has placed in front of scientific and governmental authorities over the last two decades, that they are still reluctant to move forward to at least consider the advisory system he proposes. Simply mind-boggling."

—Rob Mc Connell, *The 'X' Zone Radio Show*, 2011

"What's wrong with trying to warn people? How does this hurt anyone?"

—Wendy Snyder, WGN, Chicago, August 7, 2014

"We're not interested in anyone saying the sky's falling, but David Nabhan is a rational man who has studied earthquakes seriously and has intelligent, useful information to offer."

—Whitley Strieber, *Dreamland Radio*, 2011

"An intriguing and fascinating book. Nabhan's simply asking us to look at the data, the way science used to be done. Here's the evidence, you look at it and figure out what it means."

—Dave Bowman, KFIV, Modesto, June 20, 2014

"David Nabhan is no stranger to controversy for his belief that we can predict earthquakes to some degree. His theory is pinned to gravitational tides, and these are immense forces!"

—Faune Riggin, Morning News, KZIM & KSIM, Southeastern Missouri, 2013

"This is a no-brainer for me; facts are facts."

—Rick Wiles, *Trunews*, June 10, 2014

"His ideas warrant better scrutiny, because maybe there is a grain of truth in his hypotheses and all of us deserve the best chance to survive the next Big One."

—David Fox, *Anchorage Press*, August 2015

"This is the multi-billion dollar question. Does David Nabhan have the answer?"

—David Page, *KSRO Morning News*,
San Francisco Bay Area, 2011

To my family,
and to my friends on the West Coast,
and to the teachers and students of Los Angeles.

Contents

Foreword

This remarkable book challenges some important views that most educated people, both laymen and scholars, tend to share regarding contemporary science, daring to question the dogmatism that resides in our education, research, and social systems and which shapes our politics and culture. We've been told constantly by the experts, for example, that earthquakes cannot be predicted, and that therefore science cannot help us plan for the unavoidable event when seismic phenomena should disrupt our daily lives. The scientific community, however, often shies away from the social responsibility that is incumbent upon them regarding not only seismic forecasting but other controversial topics as well, notwithstanding the massive investments in research funded with taxpayers' money. Regarding some topics, it's almost as if ignorance is the preferred option.

Lest there be any doubt, and worse than not knowing, there is still the tendency in science from time to time to turn away from the scientific method and to pounce precipitously either for or against new ideas, often with deleterious results. As a professor of the history of science, I'm only too well aware of the numerous debates in science addressed in this book that went terribly wrong owing to scientists neglecting to adhere to their own code: freedom of inquiry. The ill-fated slogans that are inextricably interwoven with those failures of the past were shouted with great conviction: manned flight is impossible, Earth does not rotate, the continents do not drift, jet streams are fantastical, and so many more. Seismic forecasting also was until very recently a prominent item among that list, giving proof that after so many centuries of lessons modern science can still at times embrace the same kind of self-assurance that was visited upon Galileo in the sixteenth century. This in itself, aside from earthquake prediction, is an incredibly important problem that the book attempts to assess, and which invites us all to evaluate. After all, science was born when the confluence of Greek logic and mathematics, empiricism, and Christianity merged into the marketplace of ideas and ideological conflict, and it still may bear vestiges of the denominations and orders with which it rose up in the sixteenth century and consolidated its power in the seventeenth. Scientific reform was necessary then and should be ongoing now if we are to inoculate ourselves from doctrinaire thinking that should otherwise pervade our scientific communities.

The eminent scientists who are courageously attempting to solve the problem of earthquake prediction certainly deserve to become better known to the general public. There are few things as important as a seismic warning system for the US West Coast. Indeed, many hundreds of millions of people—in Japan, India, Italy, Russia, Chile, and many other places—would be just as concerned to see progress of this kind made in their part of the world as well. This is an enormous question—to include the politics of issuing alerts and warnings along with everything else—and one which will have equally colossal consequences. But recent history has taught important lessons in this regard; the L'Aquila earthquake which struck in central Italy in 2009 resounded with dramatic worldwide consequences.

It isn't possible to enter into the intricacies of the legal basis for the L'Aquila case here, but the significance of the proceedings addresses a culture grappling with dogma, corruption, risk, and social responsibility. The legal case filed by many relatives of the earthquake victims was countered by scientists who defended themselves by repeating the supposedly common wisdom that earthquakes cannot be predicted, and that therefore they were not responsible for anything. The judge hearing the case, Antonio Di Pietro, a renowned reformer and anti-corruption figure in Italy, noted that their communications implied differently, and that in point of fact reassuring messages were circulated in the days before the event that created confusion and a false sense of safety in the population. Further, he found that there was no a priori justification for the dismissal of Giampaolo Giuliani's research and earthquake warning. Guiliani, an independent researcher based in L'Aquila, had long worked on the possibility of integrating radon gas emissions into a system of earthquake forecasting. His alerts were ignored, and on April 6, 2009, two hundred and sixty people in central Italy died in the earthquake that transpired nonetheless.

That's what science should be all about, finding out whether ideas stand up to scrutiny or not. The L'Aquila earthquake affair, and more than a few other chapters in history referenced in *Earthquake Prediction: Dawn of the New Seismology,* illustrate that contemporary science can still generate another "Galileo case"—even such to create international press.

There is nothing worse than dogmatism in research. David Nabhan's book casts a new light on this age-old problem.

Paolo Palmieri
Associate Professor,
Department of History and Philosophy of Science
University of Pittsburgh

Introduction

We will probably never know for certain exactly how the Minoan civilization (2000 BC–1450 BC), one of the great cultures to flourish on Earth, finally came to an end. Eminent historian Will Durant declared that their culture, based in Crete and other Aegean islands, was nothing less than "the first link in the European chain." What seems certain, however, is that a series of cataclysmic seismic disasters, culminating in the most famous hammer blow ever delivered in that part of the world around 1645 BC, was the coup de grace from which the Minoans could not recuperate. Whether the titanic explosion of the Santorini volcano, which blew half of the island into the stratosphere and sent a tsunami of stupendous height to devastate the northern coast of Crete, is the kernel for the legend of Atlantis or not, what is certain is that the Minoans were knocked down so hard that they never really regained their feet again. It has been postulated that a combination of a lost fleet,wrecked harbors and coastal cities, loss of trade, and other cascading calamities weakened the Minoans to such an extent that they finally were unable to defend themselves from neighboring freebooters and enemies who took advantage of the change of fortune.

History can't spend too much time mourning the Minoans, however. The particular means of their demise may not be the exception; it might be a rule. There is growing speculation that the number of civilizations that have been rent asunder by earthquake, volcano, and/or tsunami at just the critical moment in their history—pulling down defensive walls, changing the course of rivers, increasing vulnerability to attack, etc.—is no small list. Indeed, the catalogue of potential victims includes some world-class candidates: the Mayan city-states of Quirigua, Xunantunich, and others in Central America; Megiddo and Jericho in the Middle East; the Harappan civilization of the Indus Valley. There is even question now as to whether Troy was brought down not so much by the valor of Achilles and Agamemnon but rather more by Poseidon, god of the sea, and of—earthquakes. It may well be that a few chronicles over the last five thousand years of recorded history could be in need of some revisions after all.

Even when ancient historians have given us obvious clues as to how incredibly powerful tectonic forces may have changed the course of wars or caused crops to fail and ushered in starvation, on the most well-documented occasion it took fifteen hundred years to simply take them at their word. The Byzantine historian Procopius, for example, in his chronicle of the Eastern Roman Empire's war against the Vandals in 536 AD, writes that, "during this year a most dread portent took place. For the sun gave forth its light without brightness . . . and it seemed exceedingly like the sun in eclipse, for the beams it shed were not clear." Of course, no one ever dreamed that Procopius might have been telling the truth, since a year without sunshine would usher in what we moderns would now categorize as a "nuclear winter." That is what much of the human race might have had to endure in AD 536, owing not to thermonuclear devices but to something even more powerful: volcanism. Whether it was Krakatoa, Rabaul, or Mount Tambora in South Asia, or the Ilopango in Central America, or a combination there of, it is now conjectured that some series of massive explosions may have ejected enough dust and debris into the atmosphere sufficient to bring on a "volcanic winter." Whatever was transpiring in the darkness falling on the battle lines between the Byzantines and the Vandals, however, at the same time the Irish Annals were recording a "Year without Bread," snow was documented falling in China in August, crops were failing in Indochina, and a severe drought in Peru was punishing the Moche culture.

We Americans, safe and secure in a seemingly invulnerable, indestructible, twenty-first-century cocoon of technological invincibility and surety, tend to look back on these footnotes in history as perhaps interesting but not really germane to our own lives. Not many imagine that anything comparable to a "Year without Bread" could ever happen now.

But it could.

Ticking beneath our own homeland is a colossal time bomb capable of unleashing astonishingly destructive power. The Yellowstone Caldera in the northwest corner of Wyoming is a supervolcano. It's called "super" because every so often when it empties its magma chamber it affects the entire North American continent—and the world. The last eruption—640,000 years ago—hurled 240 cubic miles of rock, dust, and cinders into the sky. When it next erupts it will likely bury much of the western half of North America in a layer of volcanic ash. All that very well could put a dent in deliveries of baked goods to more than a few people. According to most scientists' calculations, by the way, it's overdue.

Far, far more probable and much, much nearer on the horizon is an event that most people holding this book will likely see—the next "Big One" in Southern California. It won't change history, and our nation will endure. But it will be a world-class event. There will be no person in the United States who won't feel some ripple when Los Angeles is wracked with the magnitude 8.0± earthquake that cannot be too far into the future.

For these and other reasons, it's only human nature to attempt to put these matters out of the mind. And for the last century it's also why and how the "earthquake prediction is impossible" mantra, born in an age of science when such an unsupported motto had to be bandied about via telegraph, has managed to find anyone to give it credence in an era when our spacecrafts are piercing the heliopause, our geneticists have catalogued the entire three billion base pairs of the human genome, when physicians routinely transplant hearts, livers, kidneys, and lungs. It's an absurdity on its very face to proclaim that only seismic forecasting will forever remain in the realm of the impossible, in perpetuity, when thousands of things exponentially more mind-boggling have already taken place in front of every pair of eyes on Earth. It takes courage, though, to face up to the daunting task of attempting to put those words—so easy to say—into practice. It requires the most remarkable men and women to set themselves the challenging task of attempting to find real answers for the fifty million Americans at seismic risk on the US West Coast who should gratefully appreciate their efforts.

You will meet those people within these pages: an extraordinary cast of world-class scientists from a half-dozen nations—principal investigators at NASA, lead scientists at the SETI Institute, chief seismologists at the Bhabha Atomic Research Center in Mumbai, esteemed professors at La Sapienza University in Rome, physicist laureates at the Andrea Bina Seismic Observatory—all of whom have previously felt constrained to say little regarding the self-fulfilling prophecies of defeatism that have dogged earthquake prediction for so long, and are now finally, and rapidly, being ushered into history's dustpan. Their long-restrained comments will be aired for the first time within this work. The reader will be quite stunned not only by who these individuals are but by the tone of what they have to say—including their view of how the lack of open-minded collaboration may be an obstacle to creating a rudimentary seismic warning system for the US West Coast.

The Minoans could have used something like that; we'd be very foolish to look askance at it ourselves.

David Nabhan, December 2016

Explosive eruption from Stromboli volcano, Italy, December 1969

Chapter One

Shiva, the Destroyer

Ancient peoples' beliefs about the world were reflected in their philosophies and religions. Icons and ideas generated at least three hundred years before the birth of Christ conceived the opposing yet inseparable yin and yang of Chinese philosophy. They are also expressed in the dual nature of the great Hindu god, Shiva—deity of both destruction *and* creation. India and China are not only home to these great traditions but occupy some of the most seismically active real estate on planet Earth. The earthquakes that have wracked that part of the world for eons cannot be classified as simply destructive or creative, as either good or bad, as a blessing or a curse—they are *both*.

There have been a number of great extinction events in the history of the planet, moments in time when life was pushed to the edge and almost beyond. The most well-known of these dire episodes brought an end to the age of dinosaurs, at the boundary between the Cretaceous and Triassic geologic periods—the K-T Mass Extinction (now officially termed the Cretaceous-Paleogene Extinction Event). Whether it was a comet or meteor, an object from space the size of Mt. Everest slammed into Earth some sixty-six million years ago, moving at close to thirty kilometers per second. The Chicxulub Crater, marking its impact and now buried under the Yucatan Peninsula and the Gulf of Mexico, is enormous—110 miles in diameter and twelve miles deep. Until recently it was thought that the impact alone and its immediate consequences (hurling tremendous amounts of debris into the atmosphere to block sunlight worldwide, changing the environment, killing plant-life, and inducing world-wide starvation) would have possessed sufficient clout to dethrone the titans who had ruled Earth for the prior 165 million years. That opinion has changed as of late. It seems much more likely that the dinosaurs were subjected to a savage double punch, a one-two knockout blow which even the likes of *Tyrannosaurus rex* couldn't withstand. The real killer of the dinosaurs might have been *seismic*.

If nature abhors a vacuum, science detests coincidences, and sixty-six million years ago one of the greatest of them all was taking place: *two* global killing events occurring simultaneously. Something caused the crust of Earth to open—in India—and spew forth a mantle plume, a calamity thankfully rare in the extreme. A mantle plume is a colossal upwelling of abnormally hot rock that melts its way through the crust and discharges lava in volumes scarcely imaginable. A team of geophysicists from the University of Berkeley in 2015 published the newest and most accurate dates for the startlingly close timing of the asteroid impact and a particularly Homeric series of eruptions at a place known as the Deccan Traps. These eruptions ultimately covered 1.5 million square kilometers of the surface with lava. Untold trillions of tons of poisonous gases and noxious fumes accompanied this catastrophe for life. The newest evidence suggests that if the dinosaurs were knocked down by a meteor from space, the unlucky survivors were asphyxiated.

Geologists have a plausible connection between incredible blows striking from space and the resultant extrusion of molten death from the bowels of the planet: earthquakes. Certainly, the six-mile-wide asteroid that slammed into the Earth at 150 times the speed of a jetliner, energetically equal to a billion Hiroshima-sized explosions, caused earthquakes like none ever felt since. The cosmic punch would have rung the entire world like a bell. Dr. Michael Manga, a co-author of the Berkeley study, wrote that there would have been monumental tremors, everywhere, all over the planet, all at once. No one knows what causes mantle plumes but there are very credible hypotheses that wracking the entire world with magnitude 9.0 or greater earthquakes—everywhere—might do the trick. Team leader Mark Richards remarked that though the Deccan Traps had been erupting before the impact, weathering on terraces indicates that there had been a quiescent period prior to the Chixculub event and that the resulting seismic activity of the impact might have "changed the plumbing" and restarted the process, yet in a much more intensified way, accounting for the bulk of the eruptions. "This was an existing massive volcanic system that had been there probably several million years, and the impact gave this thing a shake and it mobilized a huge amount of magma over a short amount of time."

It's been posited that a similar asteroid strike and the resultant monumental mega-quakes may have preceded the opening of the Siberian Traps as well, causing a "Great Dying" unlike any before or since—the Permian Extinction, in which 90–95% of all species were killed. If an asteroid did strike Earth 250 million years ago—as has been suggested by evidence beneath the ice in Antarctica (Wilkes Land Crater) and in the Bedout High off the coast of Australia—it might have put into motion the same geologic disaster in Siberia, this one dwarfing the lava flows in India mentioned above. Here is an event lasting a million years that paved *seven million square kilometers* of Earth's surface with lava and came within a hair's breadth of terminating all life, pumping perhaps a billion tons of methane and four billion tons of sulfur dioxide into the atmosphere. All of the above, therefore, certainly makes sense of the "destruction" part of the Shiva-earthquake analogy. What about "creation," however?

Dr. Eric Force is an esteemed professor of geosciences at the University of Arizona at Tucson and a retired United States Geological Survey (USGS) geologist. Professor Force has a unique and hard-to-dispute theory about how our civilized world came to form in its incipient stages. His contention is easily seen, for all Dr. Force has done is lay out two maps, one upon the other. The first is a chart of Earth's tectonic plates and faults; the second is a map of ancient civilizations. They are almost identical. His hypothesis is that consciously or otherwise human settlement has favored fault zones due to the benefits provided by volcanic ash for farming, the up welling of use ful materials to the surface, and ample water supplies that are usually found at plate boundaries. Professor Force's apt observation connects seismicity with the rise of civilization, while the true nexus is actually far more important, exponentially more essential for the very existence of life on Earth. For when earthquakes cease to undulate across the surface of our planet, that will be the death knell for every living creature, barring none.

Earth's continents, the very platform for all land-based life, owe their existence to seismicity and volcanism. They ride on massive broken chunks of Earth's surface called tectonic plates and those plates in turn are impelled by the currents produced in the quasi-liquid ocean of semi-molten rock far below. The surface of our planet is not "terra firma"—it moves. Every year, every day, every moment it is in motion. Wrapped like a thin skin around the seething mass of magma that constitutes by far the bulk of our planet, the crust is twisted and contorted by titanic forces far below the surface. Superheated magma rises from Earth's core and as it cools nearer to the surface it is subducted again, providing the motive force for the ceaseless tectonic treadmill, pulling back down into Earth's mantle the heavier basaltic oceanic floors. This process is very much like the dynamics of the currents in the oceans or the atmosphere, except that the subterranean flow is composed of partially melted and plasticized solids and happens in slow motion. Since continental crust is composed primarily of lighter, silicon-rich, granitic materials, its buoyancy resists subduction and—thankfully for us—avoids the destructive fate of the ocean floors. Earth's continents, and especially the amazingly durable three dozen or so cratons (the stable interior portions made up of ancient crystalline basement rock) from which they are composed, are particularly long-lived. They've been locking together, separating, and recombining over and over again for eons, drifting plates forcing adjoining plates to either slide over, slide under, or slide out of their way. Where plates meet and jostle for position is where colossal forces build up and explode; earthquakes are produced at these contested boundaries, called "faults." It is only the positions of these deep fissures in the crust of Earth that have changed over the billions of years, being traced and then erased over the ages but one of the indelible signs of a planet with the requisite vigor and power to bear life.

The scorching heat of the planet's core—almost as hot as the surface of the Sun—has been driving this process since the formation of Earth, four and a half billion years ago. In all that time the core has cooled by only about 5%. Heat generated by the radioactive decay of the stores of uranium, thorium, potassium-40, and other elements below the surface acts to slow down the cooling process. Since some of these isotopes have half-lives measured in the billions of years, there's no imminent change in temperature on the horizon for the interior of the planet. Earth is geologically alive and will remain so into the foreseeable future. Without these seismic and volcanic processes to recycle vital materials and support complex chemical cycles, our planet would be converted from a fecund, verdant biosphere into a barren, dead rock.

When that tectonic engine finally ceases to operate, when earthquakes and volcanoes are a feature of Earth's past, all life too will be expunged from the story. One need not detail the minutiae of the thousand catastrophes that would assail every form of life, since the greatest disaster is one that requires very little elaboration: the atmosphere and oceans will disappear. The Sun assails our biosphere constantly, blasting away atmospheric gases into space via the solar wind, but the planet is protected thanks to tectonics. Earth's super-heated and spinning core generates a magnetic shield that minimizes the damage caused by the solar wind while volcanoes and biotic outgassing replace that which is lost. When those defenses vanish so does any chance for life. Once atmospheric pressure is nonexistent above the oceans, water will evaporate away at extremely reduced temperatures, be blown off into space, and Earth at end will someday look very much like Mars.

There's every reason, then, for our understandably intense interest in earthquakes and to revisit the age-old question as to whether or not they might be forecast in some way. The dichotomy is that while we fear them, we need them. They bring buildings crashing to the ground, but our cement never would have existed but for those very temblors in the first place. Just as every breath of oxygen we take actually kills us—a little—so too is mankind inextricably intertwined with a force far, far greater than is easily fathomed, and which is terribly dangerous and yet bountifully life-giving at the same time.

Santorini crater, Greece; remnants of Minoan Eruption, circa 1630 BC (NASA)

Chapter Two

The Ancient Enigma

The birthplace of mankind in East Africa was anything but a quiet nursery. Our ancestors evolved in the Rift Valley where a colossal slab of Earth's surface, the African Plate, is splitting into two giant pieces. Australopithecines would have felt the ground moving under their feet thousands of times in the last two million years and would have gazed up as many times to witness Mt. Kilimanjaro, Mt. Kenya, and other now dormant volcanoes exploding. The first pre-humans shared this valley with a long list of terrifying predators but it is certainly plausible speculation to presume that earthquakes and volcanoes would have caused a different kind of fear and wonder. As Dr. Eric Force has pointed out, Neolithic farmers whose ancestors had moved out of the Rift Valley in Africa to Europe, the Near East, and southern Asia traded one danger zone for others, as great fault lines run beneath almost all the areas of early human civilization. It wasn't until the invention of writing around 3,000 BC that people in ancient societies left their first ideas about how they viewed earthquakes.

It was the Egyptian god of Earth, Geb, who made the ground shake when he laughed, according to the mythology. In Japan it was the fault of a giant catfish named Namazu, who lived at the center of Earth. The god Kashima wrestled this beast and pinned him against a monstrous boulder. However, every once in a while, as Kashima's mind wandered off, Namazu would seize the opportunity to start squirming again. In India it was thought that a column of eight giant elephants upheld the platform of Earth. Humans above would feel it when sooner or later one of these beasts would tire and put its head down. For New Zealand's Maori, the shaking was explained in oral traditions by the kicking in the womb of Mother Earth by her next unborn infant.

The Mayans and Aztecs had earthquake gods, Cizin and Tepeyollotl respectively, and their societies, along with those of other pre-Columbian Mesoamerican civilizations, were fairly obsessed with earthquakes. The evidence is strewn all over the heartland of Mexico and Central America. The calendars of these peoples were so important to the Maya and other tribes because they calibrated their cycles of "world ages," all ending in cataclysmic disasters. The fourth world, the one in which the gods placed humans, came to an end on December 21, 2012—as many around the world awaited a doom that had been predicted since the calculated beginning of the "Long Count" of the ancient calendar: August 11, 3113 BC.

There are still a few enigmas about all this, though. There is no scholar on Earth who can say how old the Long Count is, but it does seem certain that the Maya didn't invent the Long Count. They inherited it from the Zapotec, and they from the Olmec. The absolute oldest physical reference we have, according to Professor

Jon Olson at California State University in Los Angeles, is to be found carved in a stela at San Jose Magote in the state of Oaxaca, Mexico. This monument dates from around 600 BC—long predating Mayan civilization—and the inscriptions were carved by Zapotec scholars. On this stela are inscribed the names of the twenty days of the Sacred Calendar as well. One of the glyphs deciphered for one of the days is laconically translated into English as "earthquake."

The names given to the days provide an excellent window into the collective mindset of the Zapotec and Maya and others. No people on Earth have chosen the very names of their days with slap-happy nonchalance. There is no society that ever had a "Left-Handed Scissors Day" or a "Burnt Toast Day" as part of their calendar. Quite to the contrary, the name days of the calendar of any people are the result of centuries of careful selection. From among the hundreds of thousands of physical things and abstract concepts available to ancient Mesoamericans as the pool from which to pick only twenty names, it is quite insightful that they chose to spend one of their precious selections on the idea of "quaking." (The Aztec calendar also has a quake day.)

In both the Old World and the New, earthquakes were considered as far beyond human reach as the divinities themselves, and the idea of forecasting them was as fantastic as predicting the mind of the gods. It's hardly possible to delineate the exact moment in history when the hazy line between superstition, lore, and religion regarding earthquakes coalesced into something approaching scientific observation. If one were forced to make the attempt, however, to point to the exact time and place when people first attempted to deduce some sort of logical precursor to a seismic event, the selection may not be that surprising: ancient Greece. In this country where science and mathematics were born is to be found another very old example of the unequalled power of Greek observation and hypothesis.

In 373 BC a terrifyingly powerful earthquake first struck down the city of Helike in the Peloponnesus on the Gulf of Corinth, and the tsunami that followed washed the destroyed city out to sea. Helike, the renowned cult center for Poseidon, god of the sea and earthquakes, was ironically completely erased from Earth by the very powers to which it paid homage. It was noticed nonetheless that just prior to the end, all manner of animals seemed to have left the city en masse. The classical Greek writer Claudius Aelianius (also known as Aelian), writing in the second century AD and quoting sources now lost, reported that "all the mice and martens and snakes and beetles and centipedes and every other creature of that kind left in a body by the road that leads to Keryneia, fleeing the coastal city for higher ground." This exodus took place over the five days prior to the disaster and was a source of amazement and mystery for the people of Helike who, according to Aelian, most certainly noted it.

Astrophysicist Thomas Gold has posited an idea that earthquakes might be predicted by noting precipitous changes in the pressure of subterranean gases. His theory is that temblors might be triggered by the release of gas that had previously been instrumental in acting to clamp fault lines together under pressure. To his mind, it seems plausible that the ancient inhabitants of Helike hadn't falsified the account of the destruction of their city and that it might have been the case that animals with far more sensitive senses of smell could have been alerted to the impending disaster if there had been outgassing before the catastrophe. Gold put his hypothesis in front of geoarchaeologist Steven Soter, a friend and colleague at the Cornell University Center for Radiophysics and Space Research. Soter was intrigued by the idea and, with Greek archaeologist Dora Katsonopoulou, inaugurated the Helike Project. They used all the clues at their disposal, both ancient and modern, and hit pay dirt in 2001, when they excavated what are undoubtedly the remnants of the fabled city.

On December 26, 2004, only a few years after Helike's discovery yet two and half millennia after the incident, an equally remarkable demonstration of animal perception of impending seismicity took place. The Indian Ocean quake and tsunami that killed a quarter of a million people, numerous news agencies came to realize, strangely enough killed too few elephants, buffalo, and other animals to even count. They too had all, in Aelian's words, "fled for higher ground." The National Geographic News was on the scene this time, among others, to record the event, a source most would deem at least equally reliable to Aelian.

The implacably practical Romans might have paid lip service to the idea that Neptune's anger caused earthquakes. Many high-born, well-educated patricians, though, had left that and other superstitions for the proletariat. Moreover, the same martial character that caused them to waste little time worrying about defeat but instead to focus on sizing up an enemy and finding the means to bring about victory served them here too. If it was the Greeks who initiated the first tentative hypotheses, it was most certainly not in our modern age that the first words were uttered about earthquakes as purely the phenomena of natural science—but in Rome, five decades before Jesus Christ was born.

Lucretius (99 BC–51 BC) in Book VI of his classic *De Rerum Natura* describes the causes of earthquakes in such modern terms and so completely free of hocus-pocus that one has to glance back every once in a while at the imprimatur to ascertain that the work was indeed written by a contemporary of Julius Caesar.

"Now attend and learn what is the reason for earthquakes. The Earth bears many lakes and many pools and cliffs and caverns in her bosom. And many rivers hidden beneath the Earth roll their waves on submerged stones. For since clear fact demands that it should be in all parts like itself, she then has these things attached below her and ranged beneath. The upper Earth, then, trembles at the shock of some great collapse, when time undermines huge caverns beneath. For whole mountains then fall, the trembling going out abroad from the place."[1]

Lucretius wasn't the only noble Roman who thought there were better places to look for the causes of earthquakes than Mount Olympus. Seneca (4 BC–AD 65), in *Quaestiones Naturales*, in writing about a violent earthquake that had struck Pompeii in AD 62 is perplexed enough to say that "the thread of my work, and the concurrence of the disaster at this time, requires that we discuss the causes of these earthquakes." He notes the timing of the strong shaker: "The disaster happened in winter, a period which our forefathers used to claim immunity from such dangers." He makes it clear no superstition or sacrificing of chickens will be countenanced in his treatise: "Be assured that none of these things be the doings of the gods, and that the moving of

---

[1] Lucretius, *De Rerum Natura, (On the Nature of Things)*, Loeb Classical Library, Harvard University Press. Translated by W. H. D. Rouse.

heaven and earth is no work of angry deities. These phenomena have causes of their own." Two thousand years before seismologists had begun to hypothesize that great quakes might "reverberate" around the Pacific's "Ring of Fire," Seneca was postulating something very much like that. "The province of Asia lost at a single stroke twelve of its cities. Last year calamity overtook Achaia and Macedonia, now the injury has fallen upon Campania. Whatever be the nature of the force that assails us, it makes a circuit, paying a second visit to places long passed over, in some places more rarely in others more frequent." Just a few years after Seneca penned these words, Pompeii found out, in AD 79, the *real* meaning of the word "calamity."

In that year, the great naturalist Pliny the Elder (AD 23–79) was by chance just across the Bay of Naples from Mt. Vesuvius when it erupted. He had recently been appointed Prefect of the Roman Navy and was stationed a few miles away at the port city of Misenum. The equally renowned Pliny the Younger stayed at Misenum but relates how his courageous uncle ordered galleys launched to help rescue people at Herculaneum and Pompeii, while he instead "took a fast-sailing cutter," anxious to make first-hand observations of the event before it ended. Needless to say, he needn't have hurried; Vesuvius gave him all he could have wished for, and unfortunately a lot more, too (as one can discern from the end date of Pliny's life above).

As early as the first century, some people at least had come to realize that earthquakes were natural phenomena, that they must have logical causes, that those causes might be understood—and that hopefully some sign of the impending disaster could be recognized so as to ameliorate the awful damage they caused.

When the Roman Empire fell, the Catholic Church was installed as the overseer of all learning in the West. Things became much more simplified at this juncture, as all knowledge sprang from a single book. The Bible explained anything anyone needed to know, according to the way the Church translated it, of course, since it was the rare individual in those days that could even read or write his or her own name. Earthquakes were very simply and very obviously an act of God. Anyone who went looking for other reasons did so at their own risk.

An unfortunate coincidence of history in the Middle Ages cemented the idea that earthquakes and divine punishment were inextricably interwoven. On January 25, 1348, an unusually powerful earthquake centered in Italy shook the heart of Christendom. Simultaneously, the bubonic plague, which had broken out in Central Asia, had made its way to the Crimea, sweeping the steppes and depopulating vast areas. The Genoese maintained a trading presence in the Crimea at Caffa. In 1347, this Christian outpost was at war with the Mongols, who had invested the town. When the plague arrived, the besiegers became the besieged. The Mongols stubbornly maintained the blockade until their numbers were so decimated that further hostilities became pointless. In a last act of spite, they loaded their catapults with the disease-ridden bodies of their dead comrades and launched these biological weapons over the walls of Caffa. What the Mongols had been unable to do the plague accomplished with ease. The remaining Genoese left alive took to their ships and made it back to Italy, bringing with them the seed of the worst single disaster ever to befall Western civilization.

The shock wave that was the Black Death swept across the continent in 1348, in some places weeks or months after heaven's obvious death knell: the prior earthquake that had advertised God's impending wrath.[2] Whether it took out a third of Europe's population, or half as some scholars think, even the most dim-witted serf would most certainly have been able to make *this* connection. Earthquake prediction, it should go without saying, made very little progress through the rest of the fourteenth century, half a continent having been emptied after the last one striking.

It wasn't until the Age of the Enlightenment that another seismic disaster called into open debate the opposing views of nature versus divine retribution. On November 1, 1755, Portugal's capital city was shaken to its core by a very strong earthquake and, after being thrown down, pummeled by the resulting tsunami. The casualties were horrific. Historian Alvaro Pereira has estimated that thirty thousand to forty thousand were killed and a shocking 85% of Lisbon's buildings destroyed. (A number of historians count the death toll much higher, perhaps as great as one hundred thousand fatalities). It didn't take long for the dazed survivors to start circulating the rumors, stories that spread like wildfire among the two hundred thousand homeless in Lisbon's rubble-strewn streets.

The earthquake had taken place on All Soul's Day and this in itself put many a morbid thought in people's minds and on their tongues. Then the news came that a nun, one Maria Joanna, had

---

[2] Christian Rohr, *Environment and History*, May 2003, pg, 127–149, "Man and Natural Disaster in the Late Middle Ages: The Earthquake in Carinthia and Northern Italy, Jan 25, 1348, and its Perception."

foretold the event, the impeding calamity revealed to her in a vision of Jesus Christ himself. If this weren't enough, there was more, much more. In 1752 an unknown prophet had predicted a terrible event on the coming All Soul's Day. When nothing happened the premonition was repeated for 1753, and then again for 1754, and finally, the third time being the charm, for 1755. The shock was so intense that on the anniversary of the great earthquake in 1756, the authorities had to issue a proclamation forbidding residents to leave the city, with so many thousands desperate to flee Lisbon and initiate a renewed panic. With all this, people started ranting that *someone* had to be to blame, since no more obvious sign of God's displeasure could be imagined. When the crowds started to come to the agreement that that "someone" must be King Jose himself, it was difficult indeed to dissuade them.

Oliver Wendell Holmes relates in *First of November, Earthquake Day*, that "the Portuguese believed that God's wrath had smitten their city. King Jose, huddled under a tent with his family, was persecuted by the priests who told him that this disaster was the result of his own many sins. He was compelled to lead a penitential procession with his queen and daughters, all barefoot."

This all took place, though, in the Age of Reason, and the philosopher Jean-Jacques Rousseau and others took quite a different view. Rousseau pointed out in a famous letter to Voltaire that perhaps cramming hundreds of cheaply constructed, multistoried tenements into a tiny quarter of the city had more to do with loss of life than any sin King Jose or anyone else might have committed. He also blamed the people of Lisbon themselves, for great numbers of people refused to abandon their premises at the first shock, fearing more for their possessions than for their lives—and so lost both.

The Marquez de Pombal sent a questionnaire to every parish in the country to compile information about when the earthquake began, the duration of the tremors, their direction, the effects of the earthquake on the sea and springs and rivers, the height of the waves, the width of fissures in the ground, and more. It was the first scientific study of an earthquake ever conducted.

Even such a heavyweight as the German philosopher, mathematician, and scientist Immanuel Kant chimed in, giving his opinion of the reasons for the earthquake, and actually making an earthquake prediction of his own. "The ground above which we live has to be hollow and that the vaults that form it are linked together, even beneath the sea it is observed. For example, Lisbon and Iceland, which are distanced more than four hundred and a half German miles, suffered an earthquake on the same day. The ruins of Lisbon should remind us that no building should be erected along the Tagus, as the river points in the direction that naturally the earthquakes will follow."[3]

Taking a guess isn't such a bad thing. Since our planet didn't come with an owners' manual, a little trial and error is required from time to time. Kant, the author of *Critique of Pure Reason*, and one of the truly great thinkers of the eighteenth century, wasn't predisposed to throw away such a practical tool when

---

3 Silveira, Luís (translation), *Ensaios de Kant a propósito do terremoto de 1755*, Câmara Municipal de Lisboa (Lisboa, 1955).

nothing else was to hand. Sometimes the tactic works, beautifully, elegantly. A glance at the map of the fault lines around Europe and North Africa gives proof that such a "hollow" does indeed run between Lisbon and Iceland: the awesome gash in Earth that wends its way along the bottom of the Atlantic Ocean, known as the Mid Atlantic Rift.

Immanuel Kant was a man who spent his entire life within forty miles of his birthplace in Königsberg. He was an eccentric whose daily walks were so perfectly synchronized, townsfolk could set their watches by them. The only occasion on which he set off late was when he became so engrossed by reading Rousseau's *Emile* that he lost track of time, supposedly causing confusion among the shopkeepers whom he passed. Kant, who stood a bit less than five feet tall, was such a germaphobe that he often refused to

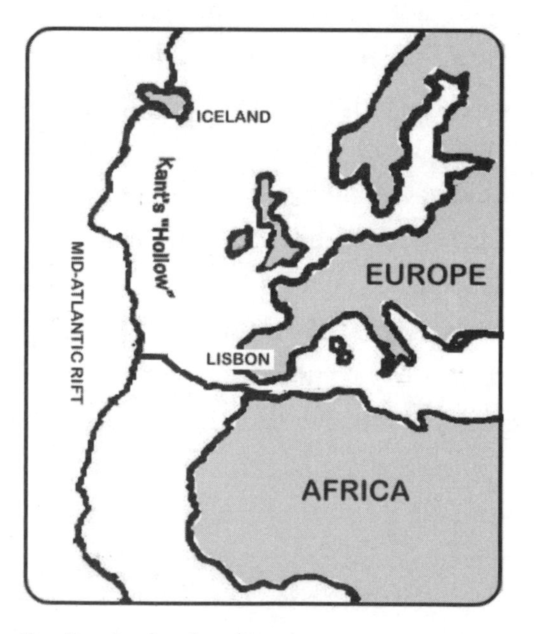

Credit: Author's collection.

acknowledge greetings for fear of opening his mouth. Nonetheless, what he lacked in stature he made up in genius. His intellect was sufficient to range 1,500 miles from East Prussia and thousands of meters below the surface of the Atlantic Ocean to presage the discovery of one of Earth's most awe-inspiring features. It wasn't until over a hundred years later when the first transatlantic cable routes were being explored that his offhanded remark was proved to be something much more.

Kant's amazingly accurate conjecture wasn't fully verified until the middle of the twentieth century, when the floor of the Atlantic was surveyed by the submarines of the US Navy's "Silent Service." This "hollow" has no equal on the planet, a section of a forty-thousand-mile-long system of ridges and rifts where molten rock is extruded from Earth's mantle to form new seabed. The new land being created, where the world is unzipped between Iceland and Lisbon, is forcing Europe and North America apart, and has been doing so for the last 200 million years, ever widening the Atlantic Ocean. Older sections of the seabed on the other sides of Earth are subducted back down into the mantle, whole sections of Earth's crust (oceanic plates) sooner or later going back whence they emerged. This also isn't too far from what Lucretius had written so long ago when he could only imagine mountains being undermined and sucked into the bowels of Earth; in fact, the truth is so much more fantastic.

Ideally, seismology should be an open-minded, adaptive, and flexible discipline. It's a new science too, not tied to unbending dogma from a past that it doesn't possess. Many are taken aback to learn that it wasn't until the 1960s that the science of seismology fully accepted the very foundation upon which it rests today: plate tectonics. Many are surprised to discover that modern seismology had to be dragged kicking and screaming to the idea of plate tectonics—forty years after Alfred Wegener delivered it to the world and only after a cacophony of scorn and titanic arm-wrestling match comparable to few in history. Seismology therefore is a very young science, a science too young to have any business using the word "never." Yet that is just how many seismologists have historically responded when the unspeakable topic of earthquake prediction has been broached. This reluctance to touch the subject makes for a very incongruous situation. It's as if aviation engineers only existed to catalog the reasons that make flying impossible, or doctors to do nothing more than chronicle the many ways people can die.

The time has long since arrived for seismology to take up the challenge that has always lingered in the background, unrecognized, uncredentialed, pseudo-scientific, ignored and snubbed—yet preordained to be the heir apparent to a discipline that has spent the last century fleeing from its own destiny.

Volcanic explosion on Jovian moon, Io (NASA Goddard Flight Center).

Chapter Three

Companion Planet

By the end of the eighteenth century, science entered into an "age of enlightenment" and demanded rigorous underpinnings for all of its pronouncements. Still, seismic activity was beyond understanding. No one at first had the slightest idea where such immense energy could be found to precipitate astounding movements of whole sections of the crust of Earth. There were few potential answers for the incredible muscle required to accomplish such feats. Two sources, however, were in plain view in the sky.

An interesting dynamic takes place as Earth whirls about the Sun in orbit. It doesn't circle the Sun as blissfully and carefree as one may suppose. On the contrary, Earth is yanked and tugged and stretched out of shape as it completes its orbit. To understand why, visualize three race cars hurtling around a circular track. One has the outside lane, one the middle lane, and the third has the inside lane. Let's imagine further that they're linked together with a steel rod that is welded to all three bumpers. Now let's stipulate that the car in the middle lane will be the pace car and the speed shall be set at 200 mph. Obviously, the car in the outside track will have to travel at a speed a bit greater than 200 mph to keep abreast of the middle car, because it has a slightly greater distance to travel than the pace car. Conversely, the inside car must travel at a speed just under 200 mph because it has the shortest track. So as the cars speed around the track in tandem and in what would seem to be perfect harmony, they are all actually traveling at different speeds.

Now apply this analogy to Earth as it orbits the Sun. The center of Earth will be our "pace car." And since the planet is almost eight thousand miles in diameter, there is more than ample room for two other "cars." On the inside track is Earth's surface, which faces the Sun. It is four thousand miles closer to the Sun than the center of the planet and is therefore moving slower than the "car" at the core of Earth. Earth's surface on the outside track, however, facing away into space, is four thousand miles further away from the Sun than the center, so that "car" is moving faster than the center. As with the linked race cars, it is seen that all three parts of Earth must move at different speeds during the journey around the Sun.

But there's a problem here! Unlike our imaginary race cars, Earth's "cars" will find it difficult to stay abreast of each other. There is another all-important calculus that now complicates things: the speed required to stay in orbit around the Sun. The mass of Earth closest to the Sun, moving slowest, will not keep pace with the speed required to keep it in orbit and it will tend to "fall into" the Sun. The mass on the far side, the side traveling the fastest, will tend to "fly off" into space away from the Sun, having the extra speed necessary for this escape. Only the center of Earth will feel perfectly at ease with both its position and speed. So, as our planet circles the Sun, it cannot do so as a perfect sphere. As it rotates, the face it shows the Sun is pulled like taffy toward the star (reaching its closest point at noon) while the opposite side of our planet is bulged outward toward space because of excessive speed (going furthest away at midnight). In other words, our planet undergoes extreme, constant, and real, physical deformations due to the tides, based both on the time of the day and date of the month.

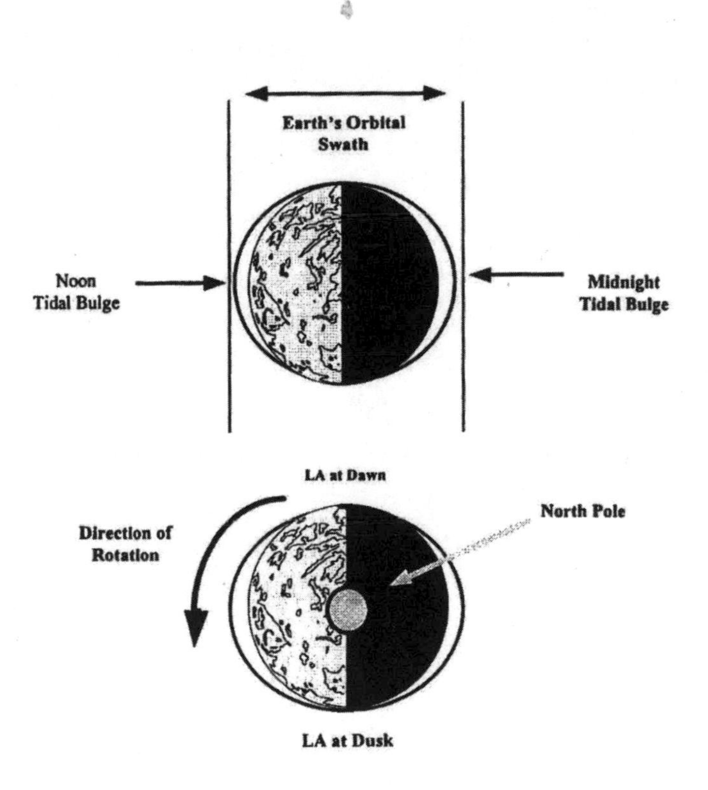

Author's collection.

Here then is the solution to the enigma that vexed natural historians and philosophers for ages—the reason for the sea's ceaseless ebb and flow, finally solved by none other than Sir Isaac Newton. While the distortions that impinge upon our planet shown in the illustration are greatly exaggerated for the purposes of demonstrating the dynamics involved, such a deformation is without question factual. It is only the degree to which such slight warping may influence changes in the immense pressures along fault lines that is at issue.

The forces just described are called "gravitational tides," and an important point to stress is that the tides are not just the movements of the ocean's waters that effect the daily changes in sea level. The tides are what *cause* this and many other unseen phenomena. Tidal forces are not only found on Earth and perceived most obviously in our seas, but are at work in any place in the universe where one large body orbits another. The larger and closer the celestial bodies are to one another, the greater the tidal effect. (In Earth's vicinity, both the Sun and Moon are very large, and very close.)

Gravitational tides are one of the primal and ubiquitous forces of the cosmos, since it is filled with larges bodies interacting with each other. The kind of energy brought to bear by tidal interaction can be unbelievably powerful, creating positively titanic forces. Some of the most stunning and iconic features of our solar system are the results of tides. The geysers of Enceladus, one of Saturn's moons, are powered by the Saturnian tides and make anything on Earth seem quite mundane. Old Faithful, in Yellowstone National Park in Wyoming, is a good contender as the most famous geyser in the world. It spews water up to heights that range between 100 and 180 feet. Neighboring Steamboat Geyser, though, also in Yellowstone, has set the record for the highest ejection ever recorded: 300 feet (ninety meters). While no one knows what the record is on Enceladus, science does have an idea of what a regular, run-of-the-mill ejection might look like. When NASA's Cassini spacecraft happened to be passing through Saturn's neighborhood in July of 2005, it took photos of explosions on Enceladus that blasted liquids and ices over eight kilometers high, ninety times higher than the best Steamboat has ever been known to muster.

Saturn's moon might impress, yet nothing can contend with the undisputed lord of the solar system: Jupiter. Everything and anything connected with this colossal planet/proto-star is over the top. That includes the Jovian tides. Heavenly bodies that orbit Jupiter must somehow contend with those mind-boggling forces. Her poor moons are pulled and squeezed and wrenched and squashed until their interiors are converted to magmatic taffy. There are no terrestrial examples to stand next to Io's volcanoes. Indeed, the ones in hell would probably pale in comparison. Suffice to say that Jupiter's innermost moon is bursting with the tallest, most ferocious volcanoes imaginable. Countless vents dot the surface, creating volcanic rings the size of Germany. The mouths of Io's volcanoes are where one would find the second hottest surface temperatures (close to 3,000° F) in the entire solar system—and yes, that *includes* the surface of the Sun. These raging monsters send furious jets of ejecta as high as five hundred kilometers above the Hadean surface.

Our planet, though, isn't pushed entirely off the stage when it comes to amazing tidal phenomena, because Earth is not only rent by solar tides, but also subjected to tidal influences twice as strong from another celestial body, one that is far smaller than the Sun but much closer: our Moon. Earth is the scene for quite a show every day with the solar and lunar tides sloshing the uncountable trillions of tons of water in her oceans around the planet with seeming effortlessness. So in searching for the source of truly titanic power, there really is no better place to look than Earth's companion planet.

Lore about full moons and eclipses is probably as old as human speech itself and comes from all parts of the world. The Moon has always been a harbinger of doom in folklore, yet the truth about our celestial neighbor is a lot more spine-tingling than any storyteller could imagine while hunkering down next to an ancient campfire.

No other planet has a satellite quite like our Moon. As satellites go, it is a monster. Out of the dozens of known planetary satellites in our solar system, it ranks fifth in size. The four that place above it, though, revolve around the huge gas giants of the outer solar system, and their mass is negligible compared to that of their companion planets. Our Moon, however, is only a third smaller than the planet Mercury. Mars, by comparison, is a little over half the size of Earth and has two moons that are barely ten miles across. The Moon is the largest satellite in the solar system relative to its companion planet, and the gravitational tides that Earth and Moon exert on each other are therefore enormous. The Sun's gravity is a force strong enough to bulge and flatten Earth's surface, but even this force is outdone by the lunar tides which are roughly twice as strong.

If current theories about the creation of the Moon are correct, its birth about four and a half billion years ago (only thirty to fifty million years after the formation of Earth itself) would have been the largest, most energetically explosive event in the history of our planet. It is believed that a body about the size of Mars (named Theia) formed in one of the Earth-Sun Trojan points, occupying the same orbital path as Earth, yet with just the proper angular distance to avoid a catastrophic impact. As it accumulated more mass, it grew large enough to surpass its orbital stability threshold, ensuring a collision with the proto-Earth. When Theia and Earth collided, the ensuing violence was powerful enough to shatter both planets, cause the iron core of Theia to sink and commingle with Earth's, and scatter debris into space, which would form the Moon. If this history is anything close to accurate, it's not only the most violent episode in Earth's history but also the most fortunate. One thing is certain: we couldn't do without our Moon.

In the distant past the lunar tides were much more spectacular than they are today. There is evidence to support the idea that as early as 3.25 billion years ago oceans had already formed and were being pulled by tides.[4] A billion years ago the Moon was only about ten thousand miles away from Earth and would have been *the* feature in the heavens, dominating the night sky and appearing twenty-three times bigger than it does today. It wouldn't have just looked impossibly powerful; it would have actually been so. No one can really say how big ocean tides were in the distant past, but there is general agreement that they were far beyond one's wildest imagination—maybe kilometers high. Whole oceans would have sloshed out of their basins and returned daily, scouring Earth like a powerful rasp. The geologic stratum from this age records the effect of this cosmic sandblasting in a layer of rubble all over Earth.

This picture of ancient Earth sounds very unpleasant, but the Moon's roughhousing turns out to have been extremely beneficent. For starters, Earth needs a large, spinning iron core to generate a powerful magnetic field around the planet. A magnetic field is the only shield life has to deflect harmful space radiation and the

---

[4] D. Eriksson, Australia National University, Canberra; Edward Simpson, Kutztown University, Pennsylvania, *Geology, Journal of the Geological Society of America*, Vol.34, #4 (December 2005): 253–256.

energetic particles hurled in our direction via the solar wind. What was once Theia's iron core is now part of Earth's, augmenting it and helping to power our electromagnetic defense. Without that iron there would be no shield—and no life.

Theia's collision with Earth was violent enough to slam the entire planet to the 23.5-degree tilt that has been maintained to the present day. That tilt gives us our seasons, which play an important part in the dynamics of evolution. But not only did the Moon bestow Earth's axial tilt, it is the only force that has kept it constant throughout the long eons of the celestial partnership. Without the Moon's interlocking and steadying gravitational clasp, our Earth would have spun wildly up and down, out of control since the beginning of its history—its axial tilt varying widely and giving rise to conditions that would spell climatic mega-disasters unending.

The Moon calmed the young Earth in another important way. In that ancient era Earth was spinning wildly, revolving at unbelievable speeds; the day was only a few hours long.[5] This excess rotational energy has been slowly, inexorably sapped from Earth and shifted to the Moon by virtue of the laws of conservation of energy and angular momentum. In the process of bleeding away Earth's surplus rotational power (again, through tidal interaction), that energy transfer to the Moon evidences itself in two ways: Earth's spinning slows on the one hand while the Moon's power to escape from Earth orbit increases. To the extent that the day has become longer, the Moon has moved further away. This process is still taking place today, exerting a braking force on Earth equal to two billion

---

[5] Dr. Gregory Ojakangas, Drury University, "Lunisolar Tidal Signatures in 1.9 Billion-year Sediments: Implications for the Lunar Orbit," *Minnesota Geological Survey*, February 2009.

horsepower, and lengthening the day by .001 seconds per century, and compelling the Moon to move further away in orbit by 3.8 centi-meters per year.[6]

As far as the mega tidal inundations are concerned, many biologists attest that this too was a boon for life. By splashing the oceans in and out of their basins, the Moon created the biggest, longest-lived "tide pool" in the history of our planet. Here was a place not really oceanic, nor totally terrestrial, but a hybrid—just the sort of twilight space best suited for evolutionary dynamics, a dynamic locale of primordial life.

Were it not for our Moon, evolution would have most certainly taken a drastically different turn. Life would have been forced to exist and evolve on a rapid-fire home planet, whirling maddeningly. Organisms would have had to develop in an environment where heat and light were doled out in snippets of a few hours, followed by equal durations of darkness. A weaker or nonexistent magnetic field would have allowed killer radiation and particles to flood to the surface. The only "season" would have been whatever extreme climate the erratic axial tilt might be producing at the time. Were life dogged enough to somehow cope with that, in short order the whole thing would have been reversed and the weather conditions dragged to the other side of the spectrum. One can only speculate what all this would have meant for the chances of life on Earth. Were it not for the Moon, we likely wouldn't be here at all.

---

[6] A. Roy and D. Clarke, *Astronomy, Structure of the Universe*, JW Arrowsmith Ltd. Bristol, UK, 1989.

Tokyo, 1923, after the Great Kanto Quake (Courtesy of USGS, George A. Lang Collection).

Chapter Four

Time and Tide

It was in the ruins of Lisbon in 1755 when the world under the tutelage of some of the greatest geniuses ever to exist—Voltaire, Rousseau, Kant—finally put to the side divine wrath as the source of earthquakes and brought to the fore a much more rational proposition: seismicity as the result of the interaction of yet unknown tremendous natural forces. But what could the source be of such stupendous energy capable of rending cities into dust, powerful enough to move entire continents off their foundations? In 1755 not only were fusion and fission two centuries distant, but the internal combustion engine was still more than a hundred years into the future. It didn't take long for many scientists to focus on the only possible force in existence capable of warping the planet, yanking massive sections of the planet one way or the other, triggering earthquakes: conjoined lunar and solar gravitational tides. Since the middle of the nineteenth century, scientists and seismologists on every continent have suggested that lunar and solar tides, acting in tandem during new and full moon phases, play an important part in understanding how earthquakes are triggered. The fact that hardly anyone would recognize their names is support for the argument that the dustpan of history contains quite a few items that probably deserve to be catalogued elsewhere. Here is a list of the greatest seismologists who might have contributed to a future breakthrough on how earthquakes might be reliably predicted, and about whom, bizarrely, scarcely more than the slightest is known.

Table 1: Peer-Reviewed Gravitational Tidal Studies (1845-2010)

Scientist	Title of Study	Journal	Date
Richard Edmonds	On the Remarkable Lunar Periodicities in Earthquakes	Edinburg New Philosophical Journal	1845
Alexis Perrey	Frequency of Earthquakes Relative to the Phase of the Moon	American Journal of Science	1876
Arthur Schuster	On Lunar and Solar Periodicities of Earthquakes	Proceedings of the Royal Society-London	1897
R.D. Oldham	Periodicities of the Tidal Forces and Earthquakes	Journal of the Asiatic Society of Bengal	1903
H.V. Gill	Some Recent Earthquake Theories	Nineteenth Century	1908
Myron Fuller	The New Madrid Earthquake	USGS Bulletin #494	1912
Charles Davison	Diurnal Periodicities of Earthquakes	Journal of Geology	1934
H.T. Stetson	Correlation of Deep-Focus Earthquakes with Lunar Hour Angle	Science	1935
Maxwell Allen	Lunar Triggering Effect on Earthquakes in Southern California	Seismological Society of America	1936
Michael Shimshoni	Evidence for Higher Seismic Activity	Geophysical Journal	1971
Dror Sadeh	Periodic Earthquakes in Alaska and Central America	Journal of Geophysical Research	1978
Leon Knopoff	Lunar-Solar Periodicities of Large Earthquakes in Southern California	Nature	1983
Elizabeth Cochran	Earth Tides Can Trigger Shallow Thrust Fault Earthquakes	Science	2002
Laurent Metivier	Evidence for Earthquake Triggering by Solid Earth Tides	Earth and Planetary Science Letters	2009
V.I. Kolvankar	Lunar Periodicities and Earthquakes	New Concepts in Global Tectonics Newsletter	2010

Author's collection.

There are literally reams of peer-reviewed abstracts from all over the world connecting the tides with the production of earthquakes. Indeed, typing "tidal triggering of earthquakes" into the search field at the NASA/Harvard Astrophysics Data System will elicit hundreds of others that should capture the attention of any interested investigator (http://adswww.harvard.edu/).

In the summer of 1996, the *Los Angeles Times* published a story about two world-renowned scientists who had accomplished the "impossible"—they had successfully forecast an earthquake. Carnegie Institution seismologist Paul G. Silver and Tokyo University professor Hiroshi Wakita had convinced the governments of China and Myanmar to evacuate a border district between those two countries, scant hours before a 7.3 quake struck on July 11, 1995, right where they said it would, right when they said it would, "averting substantial loss of life," according to the *Times*. Silver and Wakita shocked the scientific community when they made it clear they were less interested in accepting kudos and much more determined to deliver a dressing down to American seismologists for their foot-dragging and naysaying. The article was entitled "US Skepticism on Predicting Quakes Assailed." The tenor of the article was strong enough for the *Times* to editorialize that "it was relatively uncommon for scientists to be openly critical in print of an entire scientific community."

Seven years later, on October 23, 2004, the same *Los Angeles Times* published another watershed article: "Tidal Forces Can Trigger or Slow Earthquakes." The piece referenced a study, validated by being published for peer review in the prestigious journal *Science*, analyzing two thousand quakes by a consortium of scientists at UCLA and Japan. The lead author, UCLA's Elizabeth Cochran, concluded that the researchers suspected that "tidal forces speed up or slow down the timing of earthquakes on faults that are about to rupture because of other stresses." Cochran's findings were buttressed in 2009 when Laurent Metivier of Paris Diderot University published his conclusions in the journal *Earth and Planetary Science Letters*: a distinct connection between solid Earth tides and earthquakes.

By 2002, even the conservative USGS was giving amended reports to the media, although perhaps only to the slightest degree. Reuters' reporter Gina Keating asked the same USGS functionary, who in 1997 had told the KNBC News in Los Angeles that the only connection between tides and quakes was "random," if the prevailing view had changed. On the one hand it was answered, "if doing the easy things like the full moon worked we would be doing it." The scientist went on to say, though, that instead of the tides producing purely random results as previously stated, there now was "a slight correlation."

In 2006 a consortium of Italian and American seismologists, led by Carlo Doglioni at Rome's La Sapienza University, published their findings that the tidal force of the Moon was literally dragging the entire North American continental plate westward. That the tides are shown now to have sufficient power to accomplish something so monumental, while being relegated to insignificance as a force with the potential for triggering a temblor, seems counterintuitive to many people.

More recently, in 2010, Dr. Tom Jordan, Director of the Southern California Earthquake Center, in responding to AOL News' questions, plainly and honestly said that "some studies indicate that tidal forces may have an effect on smaller quakes." When AOL News pressed on about the "bigger" quakes, Dr. Jordan's response was "no evidence." Dr. Jordan's comments, though, are seen in a better light if juxtaposed with what is known about "small" and "big" quakes. According to USGS spokesperson Dr. Lucy Jones (in an interview with BBC anchor Marco Werman), the fact is that "as far as we can tell, big and small quakes start out the same way." Discounting a dynamic that can be shown to trigger small quakes, therefore, is tantamount to ignoring an incipient force for the big ones, too. Such logic would have doctors telling patients not to worry about a malignant tumor—because it was just a "small" one.

The most plain-spoken evidence however was announced only very recently—a study published in the esteemed journal *Nature Geoscience* in September 2016 sent shock waves all over the world, engendering articles entitled "Big Tides Could Trigger Large Earthquakes, Study Says" in the *Los Angeles Times* (9/12/16) and "Major Earthquakes Might be Caused by Moon" in the *Washington Post* (9/13/16). After considering everything, one is left wondering if the question of whether or not earthquake prediction via the tides is possible has gone from supposedly ridiculous to validated by science, and all in just a few short years. At this writing the profusion of studies all over the world linking tides and quakes is beyond anyone's ability to list or reference and the short answer is that this very well may be the case.

George Santayana's now oft-repeated quote was that we risk repeating history by ignoring it, but there is a real and apt example of his dictum that concerns one of the most horrific seismic disasters ever to occur. Charles Davison was one of the greatest seismologists of any era, publishing close to a hundred peer-reviewed papers in the most prestigious journals all over the world. Davison wrote *On the Diurnal Periodicities of Earthquakes* in 1896, addressing a study he made of earthquakes in Tokyo in which he observed an extraordinary aspect about those quakes. "There is in quite a few cases a marked diurnal period occurring at noon, more specifically between 10:00 am and 12:00 p.m.," Davison noted. We might want to take something written so long ago with a grain of salt, except for the judgment of history. Twenty-seven years after *Diurnal*

Periodicities was published, a cataclysm occurred in Tokyo, certainly one of the most severe earthquakes ever recorded. Approximately 145,000 people were killed, countless others were injured, and Tokyo, at that point nothing more than a smoking pile of cinders, was actually in danger of being erased from the map of Earth. Some government functionaries even suggested that the capital be moved and discussed potential sites. The shock was so great that the day after the quake berserk mobs began a week-long series of murderous rampages, killing any Korean upon whom they could lay hands. The anarchy and horror produced by such a frightful blow to society (to include towering, biblical whirlwinds of fire hundreds of feet high that ranged across the city unrelentingly, consuming thousands of people at a turn) simply caused many Japanese to temporarily and collectively take leave of their senses—letting loose an unreasoned fury on the foreigners in their midst. Perhaps as many as ten thousand Koreans perished in this fashion. The Great Kanto Quake took place in 1923, on September 1—at exactly one minute and twenty-eight seconds before noon.

Charles Davison was alive and well in 1923. One can only imagine what he must have thought about it.

Santa Monica Freeway collapses in Los Angeles during Northridge quake, 1994 (FEMA).

Chapter Five

Heresy and Orthodoxy

"The heresy of one age is the orthodoxy of the next."

—Helen Keller

After a series of great quakes striking in Los Angeles between 1987 and 1994—all exhibiting a rather stunning synchronicity—I determined that I wouldn't share Charles Davison's fate. Worse than saying nothing prior to an impending seismic assault, however, might be the opprobrium that might come from sounding an alarm and having nothing come of it. I first ran that risk and came to the attention of the news media after faxing notarized documents to CNN and others prior to the magnitude 5.4 earthquake in Seattle on May 2, 1996—the greatest seismic event in that city's history since 1965. I was nine minutes off with regard to the time. That garnered some attention, and none more than in Southern California.

In the late 1990s the "World Famous KROQ" number one morning show in Los Angeles was the Kevin and Bean Show. They had me on their air five times to talk about earthquakes. The fourth time I was invited, Kevin ambushed me. "It's really easy to say anything and then just drift away, isn't it?" he asked, none too amicably. "Why don't you put your money where your mouth is?" he dared. "Gladly," I answered. "What do you have in mind?"

He had $100 in mind. "Your next 'higher probability window' is March 13, at dawn, right? I'm willing to wager that nothing happens, and I'll bet you $100 that I'm right." My response wasn't expected. "No," I countered immediately, "let's make it $200. Deal?" He acceded to my wishes and the parameters were set. An earthquake of magnitude 5.0 or greater had to occur between 4:45 and 7:55 a.m. in Greater Los Angeles on March 13, 1998 in order for my hypothesis to gain traction and for $200 to accrue to my account; otherwise a debit of the same amount would be ceded to the Kevin and Bean Show.

March 13 in 1998 was on an unlucky Friday, so one can imagine what sort of media circus took place in the days leading up to the event. I listened to hours of people weighing in on the air about whether the "quake quack" was sane, loony, ill-informed, or . . . well . . . anything else. It was far from pleasant, but honestly, even by that early time, I was already determined to defend a line I had drawn in the sand, a line I wasn't going to back away from under any circumstances.

I appeared on the Kevin and Bean Show for the last time on Monday morning, March 16, 1998. It was in person, since I'd arrived to pick up my check. The only snag was that KROQ had already made it out to the order of "David Nabhan," and that wasn't going to do. I insisted that my winnings instead be made out to the 61st Street School's Student Fund so that the entire fourth and fifth grades of that school in South Central Los Angeles would be able to go on field trips to both the California Science Center and the Natural History Museum across from the Coliseum. There never had been a public bet like that in the history of Los Angeles—or probably any other city—and it must have been technically illegal in any event. No vice commissioner had the temerity to show up to confiscate the winnings, though. What had happened was that many citizens of Los Angeles, from every walk of life right up to the city fathers, that day came to strongly consider a message that had been rippling—like an earthquake—across the Southland.[7]

KROQ's producer, presenting the revised check, with a quizzical look on his face, gave out with this last brusque and direct question: "Who are you, and how did you do that?" This is probably a good place to answer.

I arrived in Los Angeles in the summer of 1987 and had barely settled in when I first experienced a quintessential Southern Californian phenomenon. On October 1 my alarm woke me at 7:30 a.m., and unusually for me, I hit the snooze button. Ten minutes later I was still lying under the covers, rubbing my eyes and procrastinating. Two minutes after that, at exactly 7:42, I bolted out of bed as never before. The epicenter of the Whittier Narrows earthquake was no more than a dozen miles from my apartment in Monrovia, and none of the temblors I've felt or heard since was anything like

---

[7] Technically, the bet was lost. March 13, 1998 turned out to be, as Kevin had guessed, a very quiet and nondescript day in greater Los Angeles. However, 48 hours prior, at 4:30 in the morning, at dawn, a magnitude 4.5 shaker woke up the entire town of Redlands and quite a few other communities abutting that suburban Southern Californian city. Kevin, realizing the most important aspect at stake in the wager, aware that it was the scientific method rather than a couple of hundred dollars at risk, decided that the spirit of the bet—in his mind—had been successfully carried out rather than insisting on the exact letter of the law.

this one. It came through roaring, as powerful a freight train, but with an unfamiliar noise. Outside, the eyes were treated to an extraordinary sight. The thirty-yard-long sidewalk that ran the length of the central courtyard in the apartment complex was rippling, heaving, and rolling from the street in front to the alley at back, in exactly the same fashion as one would shake out a carpet to clean it. Concrete *can* be manhandled in exactly the same way as a rug—so long as Mother Nature is doing the housecleaning. My neighbors were not only taken aback by the earthquake, but had to bear the slight discomfiture of the state of my dress, or rather undress. I was clad, but just barely, and had no thought yet of going inside my apartment for any reason. When finally a little embarrassment began to rise inside me as I realized how many parts of my apparel were missing, I was comforted by the thought that at least nothing like this could ever happen again. And it didn't—not for a few years anyway.

On the morning of June 28, 1991, at exactly 7:43 a.m.—a one-minute difference from the precise moment of the Whittier Narrows earthquake—I was attempting to sleep in again. I had been working on my sister's cabin in Sierra Madre. Like so many domiciles up in the canyon, hers was being threatened by a precipitous hillside, made worse by rains. The Sierra Madre earthquake's epicenter wasn't a dozen miles away as with Whittier though; it was directly under the cabin. Again, I was thrown out of my bed. For a second time I had to flee in a state of disarray into the street. It took me a few hours to realize that the last two earthquakes had struck sixty seconds apart. No one, though—not a single soul in Los Angeles—was saying the first word about it, and so I assumed that it must not have been anything worthy of mention.

For those who flinch at supposed unexplained and amazing occurrences, what follows next must rank among any list of jaw-droppers. The very next year, on the *precise* anniversary of the same day—June 28, 1992—not only did the next great earthquake take place, and of course at dawn, but almost as if to make matters plain beyond doubt, there were *two* of them. I was coincidentally in Big Bear when one of the greatest seismic events to ever have occurred in the continental United States transpired a few miles east of the resort town—the 7.3 Landers quake. I was fifteen miles away from the epicenter—again in a cabin, this time rented—along with my girlfriend, when it struck at 4:57 a.m. Everyone vacationing around Big Bear Lake had barely recovered from the shock when the next blow arrived. At 8:05 a.m. the ground heaved violently; a second magnitude 6.2 earthquake was occurring directly under Big Bear. It should be noted, seismologists make very plain, that the Big Bear earthquake was *not* an aftershock of Landers. These two anomalous shakers were catalogued as two separate and distinct tremors, and still are. Yet again, there was silence. Again, no one—not a single voice in the media, not one government official, not one spokesperson of the scientific community—breathed a solitary syllable about any of these obviously amazing seismic "coincidences." And so, neither did I.

The early morning hours of January 17, 1994 found me returning from a very late night. It was Martin Luther King Day, the Southland (Los Angeles, Orange, Ventura, Riverside, and San Bernardino Counties) was enjoying a three-day weekend, and even at 4:30 a.m. my girlfriend and I were just putting the final touches on a night that started out in Glendale, continued in North Hollywood, and now had us back at her home in Pasadena, not that far from Northridge, approximately twenty-five miles. So when the temblor's waves wracked the San Gabriel Valley, they were certainly strong enough. This time, however, I didn't flee into the street. Actually, if memory serves correctly, I didn't even make that much of an attempt to do anything at all. I simply placed myself over my girlfriend and, without saying a word, rode it out. When it was over and I'd calmed her down enough, the first cogent words I said to her were the words I was destined to repeat for the next two decades: "It's tidal." She didn't understand, of course, so I repeated it. "These earthquakes, they're triggered by the tides. This is a tidal phenomenon. It has to be." She *did* understand that, but from the look on her face it was obvious she hadn't the slightest interest in the epiphany, and nor did anyone else in Los Angeles, or the world. In the days after the disastrous Northridge earthquake, among the millions of words uttered and the gallons of ink expended writing about it, just as in all the previous occasions—no one suggested what now seemed clear to me.

There might be plausible reasons for such disinterest. For the entire century prior to Northridge, mainstream science had been repeating a nonstop litany regarding the certainty that nothing even approaching a pattern could possibly exist concerning how and when seismic events should evidence themselves. That an obvious design should have placed itself in front of every functioning pair of eyes of the ten million residents of Greater Los Angeles, and done so with astounding blatancy, but still managed to not elicit even the slightest questioning as to the validity of the supposed lack of patterns, is sure evidence that by repeating something often enough—whether fact or fiction—it sooner or later becomes accepted as true.

Nonetheless, there was one almost insurmountable problem: convincing ten million Southern Californians and many more tens of millions on the West Coast to summon up the courage to disagree with the civil and scientific authorities of both California and the United States. Fortunately, the US Navy and the Pasadena Public Library came to the assistance, and with allies such as these, nothing is impossible.

In 1994 one couldn't go online, push a button, and find the answer to any question on Earth. In those primitive days the only information as to the exact position of the Sun and the Moon over the decades when earthquakes had wracked Los Angeles was lodged in dust-covered annals in basements around California, in nautical almanacs and other ephemeris. In 1994, hunched over these references, I came to realize how and when the next great seismic event in Southern California would most probably occur. As Scott Cox, KERN-TV and radio Southern Californian show host, put it when he invited me onto his show in 2014, "This is the first guy to go back and check what all those seismic events have in common? The idea that it just took somebody to say 'let's look and see if there's a pattern' is pretty shocking to me."

There *was* a pattern, and here it is in plain English. The Sylmar quake hit at 6:00 a.m. The Whittier Narrows and Sierra Madre quakes struck one minute apart (7:42 a.m. and 7:43 a.m.), while the Big Bear quake (7:50 a.m.) followed them by only a few minutes. The Landers quake, the Gorman quake, and the Northridge temblor all occurred within a very tight time window—4:57 a.m., 4:51 a.m., and 4:30 a.m. respectively. Gorman and Landers, separated by forty years, nonetheless kept the same timetable: by six minutes. The pattern of quakes also striking at dusk can be evidenced going back all the way to the disastrous Long Beach shaker of 1933, which struck at 5:54 p.m.

Here we have eight of the greatest seismic events to rock Southern California over a sixty-one-year period and every single one occurred right around dawn or dusk. The earthquakes that occur on the fault line north of Los Angeles also follow the same patterns. The two greatest seismic disasters in the history of San Francisco battered that city with an interval of eighty-three years, and almost twelve hours apart: 5:12 a.m. and 5:04 p.m. (PDT). The earthquake that devastated Coalinga rolled into town in 1983, 120 miles south of the Bay Area, and only twenty-two minutes away from sharing the same destructive moment San Francisco would experience in 1989. Anchorage was laid waste in 1964 by the most powerful quake ever to occur in North America—at precisely 5:36 p.m.

But not only the time of day—indicating solar tidal influence—is germane in this investigation. It is important to note that even if the next great seismic disaster on the West Coast might occur at dawn or at dusk, it is crucial to be able to determine *which* dawn or dusk. If the solar tides establish the hour, it is the lunar tides that may decide the date. Very plainly, it is when the Moon is either full or new, the lunar tides now aligned in the same direction and pulling in tandem with the solar tides, that the combination of those two colossal powers would be most apt to trigger a seismic event on the surface of Earth below, the cosmic nudge perhaps being the final burden upon the back of a stressed fault line that may already be bursting with pressure and about to fail in any event. All the "killer quakes" in greater Los Angeles during the twentieth century took place either at dawn or at dusk. How many also bore the earmarks of having been induced by lunar tides, striking during the syzygies—when Earth, Sun, and Moon are aligned and tidal pressures at maximum—during new or full moon phases? With regard to Southern California, there is empirical evidence to suggest that while science has grudgingly come to declare that tidal triggering of earthquakes should finally be noted as at least a minimal force as an impetus for triggering seismic events worldwide, that may have nothing at all to do with what is actually happening on the US West Coast. Taking into account the data below, it seems worthwhile to weigh the possibility that there may indeed be some dynamic in play for this particular swath of the world's crust that is anomalous, marked, and striking.

There are approximately 8,766 hours in the year (leap years excepted). Highlighting a three-hour target window corresponding to local dawn and another three-hour interval to local dusk (six hours total out of the twenty-four-hour day stipulated as the durations between 11:45-14:45 UTC and 23:45-2:45 UTC) within which large earthquakes might strike gives a random probability of 25% chance of a sizeable seismic event transpiring during that time span. In the sixty-one years between the 1933 killer Long Beach earthquake and the 1994 Northridge temblor, six quakes powerful enough (magnitude 5.8 to 7.3) to have taken lives (including Long Beach and Northridge) have struck within the seventy-mile radius of the city center of Greater Los Angeles, killing a total of 262 people. Every single one struck within that thin dawn or dusk time window. Taking .25 (for the 25% chance) and raising it to the power of six gives the probability for this having occurred randomly: roughly 1 out of 5,000 (.000244). Moreover, two-thirds of those earthquakes occurred not only at dawn or at dusk, but also within thirty-six hours of the precise instant of either new or full moon phase. Hitting that meager target (approximately 5% of the time in the year) gives a number even more unlikely as far as coincidence is concerned—l in 10,000 (.000093), lending more credence to what seems a possible speculation for such an implausible historical record: gravitational tidal triggering. A table of those earthquakes is listed below, along with the mathematics used to calculate the figures.

History of Earthquakes (Mag. 5.8 to 7.3) Producing Fatalities in Greater Los Angeles: 1933-1994

Earthquake	Mag.	Solar Tides Time of Day	Lunar Tides Time of Month †	Death Toll
Long Beach 3-10-33	6.4	Dusk-5:54 PM/1:54 UT	Full-24 hours, 53 min.	115
Gorman 7-21-52	7.3	Dawn-4:51 AM*/11:51 UT	New-11 hours, 39 min	12
Sylmar 2-9-71	6.6	Dawn -6:00 AM/14:00 UT	Full-17 hours, 21 min	65
Whittier 10-1-87	5.9	Dawn -7:42 AM*/14:42 UT		8
Sierra Madre 6-28-91	5.8	Dawn -7:43 AM*/14:43 UT	Full-35 hours, 44 min	2
Northridge 1-17-94	6.7	Dawn -4:30 AM/12:30 UT		60

*Daylight Savings Time †Duration between precise instant of new/full moon phase and corresponding earthquake.
‡ $nCr \times P^r \times Q^{n-r}$ n=total earthquakes (6) r=dawn/dusk/near-syzygy quakes (4) P=probability (.0513) Q= 1-P (1-.0513=.9487) nCr= Combination of n things taken r at a time.

Author's collection.

The evidence above supports two other monumental events. The last "Big One" in Southern California, the January 9, 1857 magnitude 7.9 Fort Tejon earthquake, also struck in the very early morning (8:20 a.m.) and under a full moon (sixteen hours forty-eight minutes away from totality). Equally worthy of attention is the fact that that the largest seismic event in the recorded history of the North American continent took place at dusk, on Good Friday, March 27, 1964. This magnitude 9.2 colossus destroyed Anchorage in the strictest conformance with when the solar and lunar tides should have been at optimum if the record above speaks accurately, remarkably only forty-seven minutes away from the exact moment when the Moon entered its precise moment of fullness as far as the lunar tides are concerned. The function of a good explanation is to make the complex appear to be simple. What is posited, then, is that there is an underlying, ordered mechanism producing timed repetitions of cycles—a law of periodicities—giving answer to what transpired in Southern California over a six-decade-long period of the last century.

Downtown Anchorage after Great Alaska Earthquake, 1964 (US Army photo).

It's one thing to collate all the information about a possible pattern regarding how earthquakes struck in the past and quite another to take to a public forum to present it. There is a very simple way to determine if such a dynamic holds water or not. If the pattern exists, and is plainly seen in historic temblors, all that is required is to highlight events in the future that bear the same earmarks and note whether or not Earth shakes.

That's not such a bad idea, and actually, it's the "scientific method" that western civilization has been using for the last five centuries in order to prise from Earth her most closely guarded secrets. As long as something works that well, and for that long, unbroken, no fixing seems to be required.

San Andreas Fault, Carrizo Plain, California (Courtesy, USGS Photographic Library).

Chapter Six

Acid Test

When Galileo Galilei (1564–1642) noticed how the candelabra in the Cathedral of Pisa swung back and forth after having been pulled to the side and lit, his curiosity led him to experiment with pendulums of different weights and lengths. To his astonishment he found that using different weights on his pendulums did not affect the timing of the oscillations at all. They remained constant, as if they could fall back to Earth at the same speed. He then did something amazing, something no one had done in the two thousand years since Aristotle (384–322 BC) had pronounced how things fall: he checked it. He climbed to the top of one of Pisa's towers (not the one that leans, though) and dropped a light ball and a heavy ball at the same time. As they simultaneously crashed to the ground, so did Aristotelian science.

It's important to note that Galileo conducted his demonstration in public, in front of the townspeople, under the gazes of one and all, bidding them to use their eyes and their common sense in order to overcome the natural reluctance to dispute "The Philosopher," as Aristotle was commonly called, referencing his singular place as the font of all wisdom and truth in the sixteenth century. If earthquake prediction were to overcome equally long odds, the citizens of California and the West Coast should be called on as witnesses in just the same way. As it turned out, tens of millions all over the world were to be apprised of the message over the years—via many, many hundreds of media pieces on some of the most prestigious television, radio, newspaper and magazine outlets in existence. It won't do to detail the long list of eyebrow-raising events that have taken place on the air over the last two decades—with more than a few large earthquakes (in the magnitude 5.0 range, some as great as magnitude 6.2 and 6.3) successfully forecast within minutes or hours. Suffice to say that quite a few respected news venues—CBS News, *Popular Science, San Francisco Examiner, Los Angeles Daily News, London Daily Mail*—have seen the results, weighed this matter, and opted repeatedly to put the debate before their readers, listeners, and viewers. Many in the scientific community also have seen fit to give a fair hearing to the idea that there may indeed exist "higher probability windows" for increased seismicity on the US West Coast during the hours of dawn and dusk and within new and full moon phase dates.

The History of Geo-and-Space Sciences in Gottingen, Germany placed my thesis before none other than world-renowned physicist Dr. Marvin Herndon in 2013, requesting him as the topical editor for a five-thousand-word treatise I wrote outlining in depth some hypotheses for why the US West Coast might be singled out for such rough treatment by the tides, and giving an explanation for the astounding number of quakes at those particular hours and dates, which should point to gravitational tidal triggering. Dr. Herndon perused the piece, green-lighted it, and is the scientist who passed it along for peer review.[8] Closer to home, one of the oldest scholarly journals on the US West Coast, *San Diego Journal of History*, published another four-thousand-word paper in 2013 after submitting it to a panel of referees at the University of San Diego. It turns out there may be a quirk of geography connected with all of this, one judged feasible by both Dr. Herndon and esteemed professors at USD.

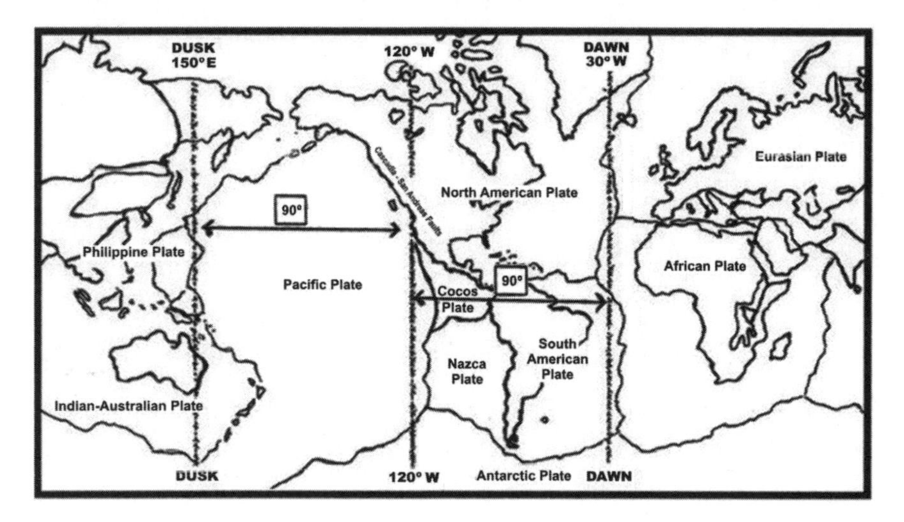

Author's collection.

A map of Earth's tectonic plates shows that dawn on the US West Coast—where the North American and Pacific plates meet—has the Sun directly above the *other* far end of the North American

---

[8] HGSS selected four highly respected professors from around the world to conduct the peer review, at the University of Athens, the University of Sydney, the University of California at Santa Barbara, and a fourth referee at the University of Pittsburgh. Three of the four professors voted for acceptance; the fourth (the Australian) voted for denial. HGSS requires unanimous acceptance, and so the piece, sadly, never ran in Germany.

plate (in the Atlantic Ocean, directly above the Mid-Atlantic Rift). Dusk in California also exhibits an anomalous parity, with the Sun directly over the other far end of the Pacific Plate (where the Philippine, Eurasian, and Pacific Plates meet). That is to say, the two plates that meet under California, Oregon, and Washington (the North American and Pacific) extend in opposite directions for almost exactly the same distance to the opposite ends of the Earth, and that those far ends are where the Sun's geographic position would be for every "dawn" and "dusk" that takes place on the North American coast.

This could be a trigger for earthquakes along the San Andreas and Cascadia fault zones in the Pacific Northwest. Twice a day Earth rotates into positions where the far plate boundaries in the Atlantic and Pacific Oceans are subjected to the strongest earth tides—and they are even stronger during new or full moon phases. Simultaneously, tidal pressures could be modulated at the opposite ends of the Pacific and North American Plates (on the US/Canadian western coasts), changing the frictional coefficient, the ease or difficulty with which the plates might slip apart, and providing the final impetus for faults already under stress and ready to fail in any event to ultimately yield. The San Andreas and Cascadia faults are almost perfectly equidistant from the far edges of the two plates in question, almost exactly in the locale best suited to receive maximum leverage from this potential tidal interaction. Further, there is yet another potent leveraging device: the weight of oceanic bulges produced. Redistribution of trillions and trillions of tons of water moved by the tides and pressing down in just the right places (or "wrong" from the point of view of West Coast residents) while lightening the oceanic burden along the California, Oregon, and

Washington shoreline might have seismic consequences. It is plausible speculation that these accidents of geography (orientation of the fault line north and south, equidistant position of the San Andreas and Cascadia to the far edges of both plates—and with those far edges also being simultaneously bulged by the tides) account for the empirical evidence that suggests that earthquakes and tidal interaction on the West Coast may have a stronger connection than is currently accepted.

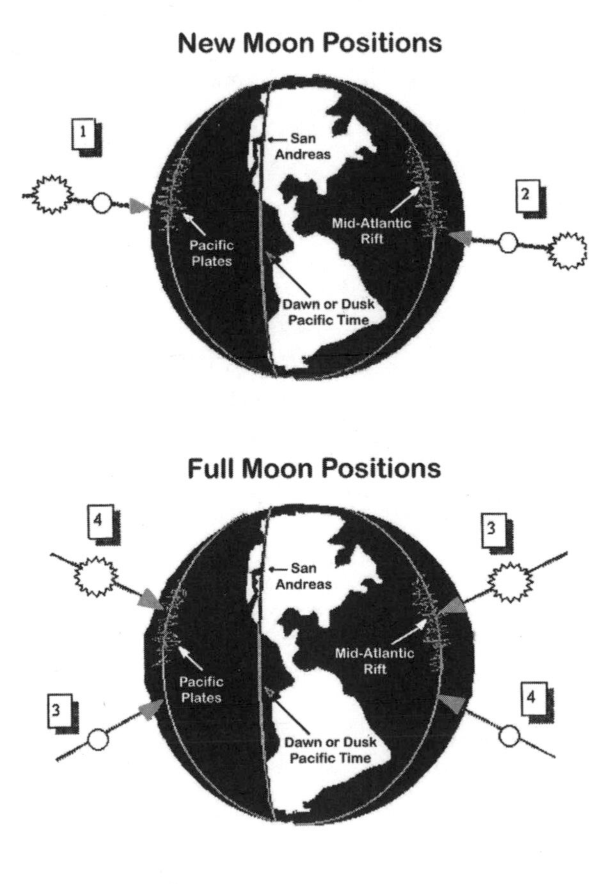

Author's collection.

That is where this matter rests at present. I have been making a public appeal for many years now that concerned citizens send their opinion to the honorable governor of the state of California, requesting that he ask for a review as to the viability or fallibility of the existence of potential higher probability seismic windows from the panel of authorities best suited to make the judgment: the California Earthquake Prediction Evaluation Council.

It's been over a century and a half since the last "Big One" rocked Los Angeles. There is every reason to suspect that the next one is on the very near horizon (more about that in chapter 12). It's not a question of "if," but of "when." Nor is this matter only of concern to Californians, for when the next great seismic event happens in the United States' second largest city, the reverberations will be felt in every corner of the globe.

Jet stream pouring moisture into California (Courtesy of NASA and NOAA).

Chapter Seven

The Shortest-Lived Profession

"Being a hero is the shortest-lived profession on Earth."
—Will Rogers

Will Rogers should have known about heroes. He was considered one himself and ironically died in the company of one of the greatest champions of the twentieth century, the first man to fly solo around the world, Wiley Post. Post first set the record for flying around the world in 1931 and was rewarded with a ticker-tape parade in New York that rivaled the one Charles Lindbergh received, earning worldwide acclaim. Two years later he circumnavigated the globe on a second occasion, this time alone. It is one of history's twists of fate that both men should have died in famous plane crashes, when Post's small craft plummeted to Earth in 1935 in Alaska with populist icon Will Rogers aboard, an accident that sent flags to half-mast all over the country. However, in the brief time between his tremendous triumph riding in the open air along the concrete canyons of New York City showered with tons of confetti and his untimely death, he received almost as many rebukes as accolades. Post had many critics and detractors, due not only to his cavalier flamboyance, but also to the fact that he discovered an extremely important scientific dynamic and refused to recant when earth scientists wouldn't have any part of acknowledging it. Wiley Post went to his death with the scientific world ignoring his obvious folly: his insistence that jet streams exist.

Wiley Post knew jet streams were real because he flew in them on a number of occasions. This barnstorming daredevil came upon them as he pushed his planes into rarefied altitudes, flying high enough to risk blacking out in attempting to set speed records by reducing air drag. Indeed, his progress across Siberia in his epoch-making flight seemed impossible to disbelievers who juxtaposed what the top speed of what his craft could have been able to achieve with the distance covered and pointed out that the two calculations didn't match. Undaunted, Post added another first to his list of accomplishments: the discoverer of jet streams was also the man who invented the first pressurized suit as a means to fly higher still.

On February 23, 1935, the *Chicago Tribune* published a news item concerning Post's brush with death in a previous crash in the Mojave Desert, only months before his demise in Alaska: "Post is Forced Down; He's Man from the Moon." It detailed how a stunned motorist came upon the flyer in the middle of nowhere, dressed in a never-before-seen "space suit." At first the driver refused to respond as Post approached, trying to wave him down and solicit his help, so certain was he that it was an alien visitor attempting to flag down a ride. "Gosh, all fishhooks, I was scared stiff!" the local man said. "I thought sure he must have dropped right out of the Moon." The article goes on, describing Post "dressed in his rubber pressure suit, an oxygen helmet over his head and face, and other paraphernalia strapped to this hobbled body, not far from resembling an apparition from Mars or a nightmare. Horror stricken, the man ran to the other side of his automobile and stood there quivering peering at the specter that haunted him." It must have taken Post some time to put the man at his ease. "Finally he got up enough courage to step from behind the car when I assured him that everything was all right and that I was human," Post is quoted as saying. A few days later, the news items about this incident were far less whimsical. An investigation into the crash had taken on a nefarious twist.

On February 27, the *Berkeley Daily Gazette* published quite a different headline: "Sabotage Seen in Failure of Wiley Post Stratosphere Hop." Palmer Nicholls, president of the Pacific Air Motive Corporation, the firm in charge of inspecting the crashed ship, declared that two pounds of iron filings had been discovered in the hydraulic lines of the wreck. So while Wiley Post may have been the toast of the town in New York City, idolized by millions around the world, it seems very plain not everyone shared that opinion. Six months later Post met his end on August 15, 1935 in Barrow, Alaska, going to his grave without seeing his meteorological discovery validated.

For the next decade, not many gave a thought to Wiley Post's chimerical "jet streams." When the final verification came, it was from the most unlikely of sources. The only scientific or governmental authorities who objectively weighed his evidence were Japanese military strategists. It was the Imperial Navy and Army who brought the Americans around to see the plain truth. From late 1944 through 1945, Japan launched thousands of airborne incendiary devices against the United States, *fusen bakudan*, wafting across the Pacific Ocean via balloons to landfall on the West Coast, riding supposedly nonexistent jet streams. Those balloon attacks, incidentally, while failing to make an important impact on the course of World War II, account for all the civilian deaths suffered by Americans on continental US soil, killing six people out for a picnic in the forests of Gearhart Mountain in Southern Oregon on May 5, 1945. (Forest rangers warn people to this day in Northern California and the Pacific Northwest not to approach strange objects they may come across in the woods. There may still be live *fusen bakudan* hidden in the brush, capable of exploding.)

Jet streams weren't the only monumental discoveries being placed on the back burner in the 1930s. As the decade began there was another terrific battle being waged in earth science. It concerned how mountains form, and the controversy would ultimately claim the life of one of the greatest geniuses of the age.

It wasn't until the 1960s that seismologists and geologists finally came to agreement as regards how it is that mountains exist on our planet. The debate about this knotty question had been intense, and as with earthquake prediction, those few dissenting voices that championed the "wrong" view were pilloried with such stinging rebukes that one is hard pressed to find comparable treatment in the history of modern science.

How mountains are created was a real enigma in science for centuries. Many a great scientist—including America's Benjamin Franklin—cracked their heads on it. Theories came and went as time passed, but after years of debate the accepted doctrine was this: Earth used to be hot. It cooled. Just as a baked apple's skin will wrinkle when removed from a hot oven and allowed to cool, so did Earth's crust wrinkle as it cooled. Those "wrinkles" are mountains. There it was—enigma solved. There was only one problem with all this, and that was that there wasn't a scintilla of truth in it. That's not how mountains are formed at all. The story of how this riddle was solved begins with a German schoolchild in the 1890s, a boy enthralled with topography, maps, and atmospheric experiments. Years later, staring at a map of South America and Africa he had received as a gift on Christmas Day in 1910, he recognized that the two continents fit together as if parts of a big jigsaw puzzle. Although Alfred Wegener wasn't the first to experience this epiphany (Francis Bacon remarked on it three hundred years prior), he was the first to act on it, changing the very foundation of seismology in the process.

Wegener's talents garnered him a position at the University of Marburg in Germany, and in 1911 he was lecturing there in the fields of astronomy and meteorology. Leafing through some research papers in the university's library in that year, he came across some intriguing information. It was a paper outlining how identical fossils could be found on both sides of the Atlantic, providing an almost perfect match between the coasts of Africa and South America. His hunch came back to him at once, and he began to scour the library for more information. It was there—all of it: geological earmarks, flora, and fauna, along with the fossil record. Everything pointed to the fact that South America and Africa had been physically connected in the distant past. And if that were true of two continents, the same must be supposed for the rest.

Wegener researched his hypothesis from 1911 to 1922. Those eleven years (interrupted by four years lost in the trenches of World War I) produced an epoch-making scientific work, *Origin of Continents and Oceans*. In this work Wegener correctly connected the diamond fields of Africa and Brazil and the coal deposits of Appalachia and England. He drew a map of Earth hundreds of millions of years in the past in which all the planet's landmass was concentrated in a gigantic supercontinent he named "Pangaea." He posited that all the present continents had been part of Pangaea and had broken away from each other and had "drifted" to their current positions over geologic time. The tremendous forces that caused them to drift also caused them to collide. By means of those awesome, continental collisions, Wegener wrote, the surface of Earth was warped and uplifted, creating mountain chains. The "baked apple" theory, according to Wegener, was half-baked.

Alfred Wegener's theory is to seismology and geology what Charles Darwin's theory is to biology. It's now called "plate tectonics" and everything we know about the dynamics of Earth science rests upon it. So it might well be imagined that Alfred Wegener received a Nobel Prize, a stipend for life from some high-minded German margrave, and congratulations from the four corners of the Earth. But things took quite a different turn.

The mainstream adherents of the "baked apple" theory went on the attack—savagely. Wegener's work was called a "stretching, twisting, contorting, misbegotten effort to fit the continents together." Wegener himself was described by American geologist Rollin Chamberin as "blind to every fact" and "a footloose type, undismayed by the awkward, ugly facts about his theory." The Royal Geographical Society dismissed it out of hand. The American Philosophical Society called it "utter, damned rot!" In New York in 1928, at a symposium of the American Association of Petroleum Geologists, he was booed off the podium and was almost physically set upon by the delegates in the audience. Professors at the University of Chicago ground Wegener's work underfoot, pronouncing that the chances of it meriting serious notice ranged from "unlikely to ludicrous." At Yale University the faculty elicited gales of laughter from undergraduates by constructing goofy, incongruous maps of the ancient Earth—a la Wegener—so that students might see first-hand that "the pieces of the puzzle not only did not fit, but weren't even parts of the same puzzle." Wegener's career didn't fare too well through all of this. He was turned down so repeatedly for professorial chairs that it became clear to him that no university in Germany would have him under any circumstances.

Wegener amazingly, refused to give up. In 1930, after being vilified by the whole scientific world for eight years, he mounted a research expedition to the ice sheet of Greenland, determined to find conclusive evidence that would link North America and Europe. Conditions were brutal; no scientific expedition had ever attempted to establish camps so far north. During one of the supply runs between his three camps, two of his companions, former students, became separated during a storm. When a head count was taken back at camp the expedition came to the horrified realization that the two young men had been left out on the ice. Alfred Wegener went out to get them—alone. He fell into a crevasse and died.

This cautionary tale is an appropriate prelude to examining what is actually known about the current state of the art regarding seismic forecasting and, more importantly, what might be discovered in the very near future about how, where, and when the next great earthquakes may strike. The renowned German philosopher Arthur Schopenhauer said that every new fact must pass through a crucible by which first it is ignored or ridiculed, then vigorously attacked, and finally accepted as though the truth had been apparent from the beginning. What an apt description of the history of science over the last five hundred years. According to dogmatists throughout the ages, Earth didn't spin, smallpox vaccines were useless, the sound barrier couldn't be broken, and the atom would never be split. We were told that if God had wanted man to fly He'd have given him wings. The idea of a Van Allen Radiation Belt was first thought nonsensical, and when the proof of its existence arrived beyond doubt, it was immediately posited that astronauts would never be able to pierce it and live to tell the tale. Anton Van Leeuwenhoek might now claim the title "Father of Microbiology," but this inventor of the microscope and the first human ever to glimpse a microscopic organism in 1676 was believed by almost no one, since Van Leeuwenhoek was considered a lowly tradesman, a craftsman. His claim that a single drop of water could be home to an entire menagerie of small animals was considered absurd. It wasn't until such a figure as Christiaan Huygens, a pedigreed scientist with worldwide stature, deigned to peer into the lens and confirmed what he saw that people began to accept the idea.

Alfred Wegener's story reminds us that every monumental idea, every great success, every stunning discovery started as nothing more than simple intuition, rough guesswork, insecure conjecture. Discoveries don't jump fully formed from just anywhere, like Athena from the mind of Zeus. Human beings are the only creatures capable of conceiving a hunch, and that germ of all genius is powerful enough to take a daydreaming young man down a path that led all the way to the ice sheets of Greenland, where, incidentally, he still rests to this day.[9] It is a road well-traveled by our ancestors, and one that has taken the human race from the interior of caves to the outskirts of the solar system.

---

9 Germany toyed with the idea of sending a naval vessel to fetch Wegener's remains in the years after his death, but became embroiled in other more pressing matters in the 1930s.

Karst topography, Li River south of Guilin, China (USGS).

Chapter Eight

Terra Exilium

"*Nulla terra exilium est sed altera patria.* No land is an exile but simply another native land."

—Seneca, *De Remediis Fortuitorum*

Giampaolo Giuliani, a technician at Italy's Gran Sasso Laboratories, made a public announcement on March 23, 2009 that the spikes in radon gas emitted from the ground that he'd been monitoring led him to believe that a sizeable seismic event might soon be on the horizon in the area around L'Aquila, Italy. In the days that followed, he may have come to regret his decision. He was reported to the police for "spreading alarm" and was ordered to take his prediction down from the Internet. Bernardo De Bernardinis, deputy head of Italy's Civil Protection Department, convened the Major Risks Committee and with six seismologists held a meeting on March 31 to reassure the public that the agency saw no reason for alarm. Unfortunately, as events were to prove, there was indeed. On April 6, at 3:32 a.m., a 6.3 quake devastated L'Aquila, killing three hundred people, injuring over a thousand, and leaving twenty-eight thousand homeless.

If the build-up to the earthquake were unusual, its aftermath was nothing less than jaw-dropping. Charges were brought against the six seismologists and the deputy head who had guessed incorrectly about the impending disaster, a trial was held, and all seven men were convicted of manslaughter and sentenced to six years in prison. Of course, the worldwide scientific community reacted with furious denunciations of the verdicts, quite understandably, and appeals were immediately filed. In November 2014, the appeals court handed down the final judgment. The six seismologists were acquitted, but Bernardo De Bernardinis's conviction, although reduced from six to two years in prison, was upheld. A great watershed had been reached in the annals of Italian science and jurisprudence. Lawyers had gone to court to protect clients using the age-old defense that earthquake prediction is impossible, and for the first time in history had actually seen it fail to win the day. As the Judge Marco Billi explained, the defendants weren't on trial for failing to predict an earthquake, but more for "contradictory and historically inaccurate statements regarding earthquake precursors" which had given the townspeople a false impression.

Italy isn't the first nation to come to such a watershed—far, far from it. Countries around the world, in almost every seismically active region on the planet, have long since instituted official federal bureaus of their governments dedicated to seismic forecasting. If the United States refuses to take steps in that direction, it risks being left behind in a crucially important scientific field by the rest of the world.

India's Ministry of Earth Sciences is considering an ambitious attempt to bolster its earthquake prediction capabilities, planning to bury the deepest network of seismic monitors yet placed in

Earth's crust—some eight kilometers below ground level at Koyna, Maharashtra—and designed to help predict earthquakes of magnitude 6.0 or greater. "We still don't as yet have good models to predict quakes," Naresh Kumar of the Wadia Institute of Himalayan Geology said of the network, "but we now have better observatory systems proposed and that should help predictions substantially." India hopes to complete the project before 2020.

In October 2012, the Russian Emergency Situations Ministry invited the US Geological Survey to consider collaborating in predicting major earthquakes. "We should pool efforts in predicting earthquakes. Russia as well as the United States has certain achievements in this sphere," Vladislav Bolov, director of the All-Russian Center for Monitoring and Forecasting Emergencies proposed. While nothing like Mr. Bolov's proposal has come to pass between the two nations, Russia has nonetheless proceeded on its own. Russians may hold the distinction of being the first major power ever to make a serious and determined effort at earthquake prediction—the enigma *ordered* to be solved by none other than Joseph Stalin, as if such matters were subject to the fiats of dictators (see chapter 9). As it is, the Emergency Situations Ministry, on its own and with no help from USGS or any other agency, issued a prediction (unsuccessful, in the end) for an earthquake to strike off the southwestern coast of Sakhalin Island in 2014.

Japan has long since put into place a governmental panel armed with the power to declare seismic emergencies. The Tokai Warning System is headed by seismologists who monitor varied indicators that are constantly scrutinized for possible precursory evidence. An event like the great Kanto Quake of 1923 was certain to repeat, and tragically did in 2011 with the devastating Tōhoku earthquake and tsunami. In the days preceding that catastrophe, which killed eighteen thousand people, seafloor instruments picked up unusual chatter and indications of slow slipping from the fault responsible for the quake. So even without being predicted beforehand outright, the Japanese remain undaunted in continuing to study tectonic harbingers. If the panel of seismologists who shoulder the awesome burden of the defense of their entire island should detect what they determine to be an ominous seismic threat, the prime minister of Japan is required by law to issue the warnings that would shut down schools, hospitals, factories, and mass transit. The board can and does examine a diverse range of data: changes in the speed and ratio of certain seismic waves, variations in the electric resistance of bedrock, abnormal geohydrochemical emissions, underground water level fluctuations, etc. In addition, a never-ending flood of information arrives in Tokyo from tilt meters, strain meters, and from seismic stations both on the mainland and offshore.

The Tokai Warning System has its own faction of critics whose disapproval has now been newly invigorated by the fact that the Tōhoku earthquake was not forecast outright, and who maintain that the money spent to maintain the whole system—$100 million per year—is "simply wasted." One of the most vocal detractors is an American associate professor of geophysics with a chair at Tokyo University, Dr. Robert Geller. In fairness to this dissenting view, in a recent article published in 2011 in the prestigious scientific journal, *Nature*, he wrote, "It beggars belief that the Japanese government operates a legally binding earthquake prediction system on this basis."

In no other country has the concept of seismic forecasting put down older, stronger roots than in China. China has an administrative branch of its government in charge of evaluating seismic danger: the State Seismological Bureau in Beijing. The SSB's Center for Analysis and Prediction has functioned as China's warning system for the last half century and seems to have earned the confidence of both the government and the public.

The American Association for the Advancement of Science published, in an article entitled "Warnings Precede Chinese Temblors (Earthquakes Predicted)" in *Science* in April of 1997, an evaluation of just how accurate the SSB has been. In the first three months of 1997, for example, seven magnitude 6.0-plus earthquakes rocked Jiashi County. Chinese officials, however, were prepared by the time the fourth struck, and the SSB "made four predictions of time and magnitude, and three were apparent successes. Their insights prompted wholesale evacuations (150,000 persons) as little as hours before the earthquakes and protected thousands of lives."

The same article then juxtaposes the American view as being both "puzzled by the Chinese events" but nonetheless insisting that "there still isn't enough information to say whether they (the Chinese) have a better understanding of the potential for earthquake prediction than we have." The Americans, while "still eager to learn more," couldn't comment further owing to the fact that "we hadn't heard that much about Chinese earthquake prediction since China opened up."

Some of the SSB's successes would have been hard to miss, being spectacular worldwide news events. For example, the SSB had been noticing ominous ground-based indicators around the area of Yingkow for some time as 1975 began. By February 4, they were convinced that a cataclysmic earthquake was about to occur. Military and civil authorities ordered the evacuation of the area at 2:00 p.m. Five and a half hours later—at 7:36 p.m.—a titanic explosion took place. The earthquake that struck was one of the strongest ever recorded, sending furious jets of water and mud twenty feet into the air, causing the sky itself to be splayed with sheets of light that have never been explained then or now. It was estimated that this quake would have killed thirty thousand people, but instead took only three hundred lives, thanks to the forecast.

The SSB makes use of a number of indicators that might be identified as precursors to earthquakes. The philosophy seems to be that there is nothing lost in considering everything, rather than risking everything by excluding something. There is hardly a precursory discipline in existence anywhere in the world that isn't also being studied in China, and that includes employing many observers in all parts of the country whose duty it is to monitor and report abnormal animal behavior. Unfortunately for many in the United States and elsewhere in the West, when "Chinese earthquake prediction" is mentioned, the eccentric image of clucking chickens, nervous cats, or barking dogs is the only one that comes to mind. It's as apt as referring to New York as "that city with all the hot dog street vendors" when in actuality there's really a bit more to the Big Apple than that.

Human evolution is framed by a millennia-long gamble of placing all its biological assets in one very powerful basket: the human brain. The human race has paid for that in other ways. We see, hear, and smell poorly. Our speed is slow, our strength weak, our toughness and healing powers relatively fragile. On the physical level there are but a handful of creatures unable to outperform us in any and every way. That's the trade-off, though, because our brainpower is expensive, leaving pocket change for every other bodily function. The average human body spends 20% of its metabolism powering an organ that comprises only 2.5% of its total weight.[10]

Other species, unsaddled with such a greedy organ as the human brain, can use that power in other ways, which are, by definition, "superhuman." There are species of snakes that see in both the visible and infrared spectrum of light. Bats and dolphins have such acute ultrasonic abilities that they can "see" with sound alone. Bloodhounds have one hundred million olfactory receptor cells, and bears' sense of smell is seven times more potent than dogs, Polar bears are documented to have found prey forty miles away by scent alone. Great white sharks can detect one drop of blood from a mile away, and many species of sharks can perceive the electric signals emitted by twitching or flexing muscles. The scientific journal *Nature* has published studies indicating that migrating birds use Earth's magnetic field to navigate from summer to winter biomes,[11] and many scientists believe the migrating monarch butterfly does so too.

The idea that everything is known about what animals can and can't do is ludicrous and can easily be disproved by the fact that something new is being discovered all the time. It's well-known, for example, that whales have the stunning ability to communicate over the mind-boggling distances between oceans through

---

[10] Richard Wrangh, *Catching Fire*, Basic Books, New York, 2009 pg. 109–110.

[11] Henrik Mouritsen, et al, in a study made at the University of Oldenburg in Germany.

infrasonic "songs." It wasn't until very recently that biologist Katy Payne discovered that elephants communicate in the same way, sending their low-pitched rumbles through the pads of their feet, into the ground, and out into the surrounding savannah.[12]

When one considers the catalog of astonishing abilities of animals, if the list were amended to include the capability to perceive imminent earthquakes, this should not come as an earth-shaking revelation. The Chinese believe that trillions of tons of heaving, flexing bedrock just might be accompanied by an attendant broadcasting of electromagnetic and/or infrasonic signals that may be perceived by some species of animals. It's not a ridiculous idea.

It is certainly understandable to put one's country and culture first, and in the case of the United States—such an astoundingly successful leader of the entire world for the last half-dozen decades at least—the natural inclination is made much stronger by the unequalled accomplishments of our nation in every field for so long. There must come times, however, when to ignore the cogent opinions of so many other great nations must at least be considered doctrinaire. In Italy, India, China, Japan, and Russia, earthquake prediction is not impossible, indicating that some introspection may actually serve us well here in the United States. The laws of physics don't change based on the time zone.

---

[12] Katy Payne, *Silent Thunder*, Penguin Books, 1998

Sylmar earthquake, Los Angeles, 1971 (USGS).

Chapter Nine

Beneath Mother Russia's Hallowed Ground

"Time has a way of showing that the most stubborn are often the most intelligent."

—Yevgeny Yevtushenko

In the early 1970s, after decades of risky encounters that brought the planet to the brink of nuclear suicide, two superpowers sanely decided that enough was enough and declared a *détente.* As part of the accord to prove good intentions on both sides, it was determined that American and Russian scientists should meet and share any knowledge that might have been secreted from one another during the long silence of the Cold War, as long as the information was not deemed strategic. Many skeptics thought that the whole thing was window dressing, and not much was expected when the scientists met in Moscow in 1971. It turned out, however, that the Russians were actually playing fairly, for one of the first things they told their American counterparts—after a quarter of a century of not speaking at all—was that they believed they had discovered some accurate earthquake prediction precursors. The Russians would be happy to present their findings to the United States if the Americans were interested. After double-checking with their translators to make sure they understood the offer and a bit of head-scratching on both sides, the Americans finally gave in, in part due to the far-sighted urgings of Lynn Sykes, one of the leading American seismologists taking part in the conferences.

The Russian method was pretty ingenious, and characteristically born of the singularly Russian instinct for refusing to knuckle under. In 1949 a magnitude 7.5 earthquake killed twenty thousand people in Ashkhabad. A few months later, another twelve thousand died in a killer quake that shattered Tadzhikistan. The Soviet government dug in its heels and made a determined decision that this continuing threat to its citizens should be met and overcome. Joseph Stalin could not believe that such catastrophic occurrences could take place without some sign being given out of the impending disaster. Mother Russia had just defeated the Wehrmacht and, flushed with her victory over the Third Reich, now took on the very planet itself. Stalin, exhibiting the unbridled hubris of the quintessential despot, and actually imagining that even earthquake prediction could be made real by the power of his fiat, ordered his seismologists to solve the problem and promptly shipped hundreds of scientists and support personnel to the hinterlands of Tadzhikistan and elsewhere. There they remained for thirteen years, completing their mission long after Stalin's death. In 1962, they proposed what has come to be known as velocity-bay or dilatancy theory. Here's what they discovered and how it works.

Every fault constantly experiences micro-quakes. They are analogous to the heartbeat of the fault. Just as a human heart can be monitored with an EKG, so did the Russian seismologists determine the status of fault lines by listening to their "heartbeats." Those micro-quakes, just like any earthquake of any size, give out a set of two basic seismic waves: P-waves (primary compression waves), which travel quicker and hence arrive at any monitoring station first, and S-waves (secondary shear waves), the slower waves that always travel behind in the wake of the P-waves.[13] The ratio between the velocities depends on the depth of the tremors. In the shallowest surface crust the differential is around 2.4 (the P-waves radiate 2.4 times faster than the S-waves). At more profound crustal depths, the ratio can narrow to 1.9. So a normal fault under normal circumstances should daily, weekly, monthly, yearly be giving out with these two kinds of waves as it produces insignificant micro-quakes, one traveling roughly two and a half times the speed of the other, depending on the depth. That's not the case, the Russians observed, with regard to faults under *abnormal* circumstances. They saw something quite different when scrutinizing unstable faults, ones about to explode.

As a fault begins to fail, the bedrock actually cracks, riddling the substratum with a lattice of micro-fissures. Now the seismic waves from the micro-quakes being monitored are no longer passing through solid rock layers, but instead through bedrock fractured with tiny cracks. These vacuoles, miniscule voids and crevices, act like baffles and moderate the respective speeds of the waves. Monitors will see the differential between P-waves and S-waves start to vary by a sizeable percentage. Then, as groundwater

---

[13] There are other kinds of seismic waves (Love Waves, Rayleigh Waves, etc.); they are not germane to the discussion here.

seeps into the cracks, the bedrock "fills" again and the velocity differential between the two waves goes back up to its normal rate. Things are far from normal, though, and water entering the cracks of a fault ready to fail is the final burden. Liquid filling the vacuoles lowers the frictional coefficient of the bedrock (making it less slip resistant) and an earthquake shortly ensues. Consequently, it should cause concern to see the background P-wave/S-wave differential on a fault abruptly change, remain at an abnormal velocity ratio for a time, and then return to normal again.

Since the Sylmar quake had just killed sixty-five people in Los Angeles in February 1971, the Americans were advised to see for themselves whether or not such a phenomenon had been occurring beneath the San Fernando Valley. Seismologists started reviewing the seismic data for earthquakes in the San Fernando Valley for the ten years prior to the 1971 earthquake. These records showed that until three and a half years before the quake, the seismic wave velocities were relatively constant. Then, suddenly, they changed sharply and then began moving slowly back toward their previous "normal" level, reaching it only a few months before the quake hit Sylmar, similar to the "dilatancy theory" results that had been found by Soviet seismologists.

In an event that foreshadowed the L'Aquila earthquake trials, in May 1976, a researcher at the California Institute of Technology, Dr. James Whitcomb, told colleagues at that highly prestigious seat of learning that he had reason to believe the same sort of phenomenon that preceded Sylmar was reoccurring. He posited that a major jolt for Southern California might be on the near horizon—within the coming year. His comments leaked to the press and he found himself repeating his prediction in front of a formal session of the California Earthquake Prediction Evaluation Council with television cameras rolling—an unusually public setting for that august body. When no great earthquake struck within the next twelve months, Dr. Whitcomb experienced some rumblings of another kind. Real estate agents in Los Angeles were furious at how his false alarm had affected the market, and they weren't alone. The harsh criticism came from city officials too, who along with the real estate brokers were threatening to sue him. Still, comrade Stalin would be pleased to know, scientists the world over continue to monitor these wave velocity variances on the great fault zones.

Whether the Russians got everything right or not, one thing is certain. The link between surface and subterranean water and cracking bedrock now seems a very important one, and the evidence doesn't come just from just Russia but from all over the globe. This opinion is buoyed by real physical evidence that the movement of water has been proven to be directly correlated with earthquakes.

The area around Lake Meade near Las Vegas experienced hundreds of tremors in the years right after Hoover Dam was built in 1935. University of Alaska seismologist Larry Gedney wrote that "the level of seismicity fluctuated in direct response to water level. None of the shocks had been particularly damaging—the largest was about magnitude 5.0—but the area had no record of being that seismically active." In the 1930s it was easy to ignore the idea that increased seismic activity might be a by-product of dam construction. And Hoover wasn't the only dam to have had its opening ceremonies punctuated with terrestrial fireworks.

The Kariba in Zimbabwe is a 125-meter-high dam that holds back a reservoir of 6,649 square kilometers of water. In December

1958 the reservoir began to fill—accompanied by twenty-two earthquakes in 1959 and fifteen in 1961, some as large as magnitude 4.0. Seismic activity kept pace with the impounding of water: sixty-three shakers logged in 1962, sixty-one more in the first seven months of 1963. Indeed, as the lake rose, the magnitude and frequency of the earthquakes increased as well. When the lake was finally completely filled in 1963, a series of over a hundred temblors wracked the area along the dam, the strongest tremors occurring in the deepest part of the lake, some in the range of magnitude 6.0 and 6.1. It is worth noting that prior to the construction of the dam the Zambezi valley, although the locale for a number of minor faults, was considered seismically quiescent, with not a single decent-sized earthquake of mention transpiring in the prior decades.

The year 1963 wasn't a good year for dams in Europe either. Seismicity is suspected to have been the reason for one of the world's most deadly dam disasters, the overtopping of the Vajont Dam in the Italian Alps in October 1963, after having been pummeled with over two hundred and fifty earthquakes. The 261-meter Vajont, the world's fourth highest dam, was completed in 1960, and as with Hoover and Kariba and so many others, as soon as the reservoir started to fill, seismic shocks were inaugurated along with landslides. In late 1960 the reservoir was partially drained and the seismic activity and slope movement ceased as well. The reservoir was then filled again, provoking a new spate of earthquakes. Heavy rains in 1963 swelled the reservoir to capacity by late summer. The dam responded immediately. In the first half of September alone sixty shocks were registered, with renewed landslides on the slopes of Mount Toc. Finally, on the night of October 9, 350 million cubic meters of rock broke off Mount Toc and plunged into the reservoir, creating a colossal wave resulting from the impact that sloshed over the top the dam by an astounding 110 meters. Two minutes later the town of Longarone was faced with this thirty-story-high wave as it raced downhill. It basically wiped out the town, killing almost every resident and leaving a death toll of 2,600 people in its wake.

A somewhat similar incident took place a few years earlier in 1959, the breaching of the Malpasset Dam above Frejus in the French Riviera. A fault below the dam is now considered the most likely cause for the disaster, which destroyed the villages of Malpasset and Bozon, killing 423 people. Time and time again, major earthquake swarms have occurred simultaneously with the construction and opening of dams all over the world: Hsinfengkiang in China in 1962; at Kremasta in Greece in 1966; at Koyna in India in 1967; at the Orville Dam in California in 1975.

The most striking evidence comes from China. On May 8, 2008, eighty thousand people were killed in a horrific 7.9 earthquake that wracked Sichuan Province. The Zipingpu Dam had just been completed, built only five hundred meters from a section of the fault line that failed, and a little over five kilometers from the killer quake's epicenter. Geophysicist Christian Klose and other scientists focused attention on such obvious connections in the esteemed journal *Science*.[14]

Dams can plausibly give birth to killer quakes in two ways. First, as the Russian seismologists noted, water is a seismic lubricant. It changes the frictional coefficient. If a massive piece of bedrock is poised to fail and held in place only by tremendous friction, water will be the final burden, and dams will force that

---

[14] *Science* Vol. 323, Issue 5912 (January 2009): 322.

friction-changing medium deep into the bowels of Earth, under pressure, to hasten forward an untimely seismic event.

But there is another way in which dams might cause earthquakes, and it's easily visualized. We have all broken a strong piece of metal by bending it back and forth repeatedly in the same place, creating a highly weakened juncture that finally snaps apart. As dams fill with tremendous quantities of water, the underlying strata of rock upon which the water pressure rests must be made to withstand the enormous increase in weight. The next season, as rains diminish or as irrigation and other uses deplete the dam's reservoir, the pressure subsides and the underlying rock will spring back. Over the years, this cycle of pressing down and springing back acts to weaken the basement under a dam, and if it is subject to faulting, if the dam is in a seismically active region to begin with, this can be a formula for disaster. Water is of great significance on this planet, with a role to play in so many dynamics of almost every process in earth science. That it should also affect seismicity should surprise no one.

Saturn, if it could somehow fit, would float in Earth's oceans despite its mass. Water is quite dense. A cube of the stuff three and a third feet on each side weighs a ton. Untold amounts of this liquid evaporate from the oceans, rain on the continents, and are cycled back again, creating the planet's rivers big and small. On the list of the big rivers, the Mississippi certainly rates mention. It's the world's twelfth largest by discharge. Every second this mighty waterway disgorges around 16,800 of those imaginary cubes into the Gulf of Mexico. If Hera had been given the opportunity to add another impossible task to the Twelve Labors of Hercules, she might have demanded that he change this river's course. Not even the vindictive queen of the gods, however, would be so merciless as to demand that its flow be stopped, or worse, reversed. She'd have realized that all the Olympian deities working together, with a few Titans thrown in for good measure, couldn't accomplish that.

Early on the morning of February 7, 1812, people along a promontory of the river where Kentucky, Missouri, and Mississippi meet awoke to find that the river was indeed running *backward*, and according to eyewitnesses, "at the speed of a fast horse." Trees were collapsing and the air was filled with sulphur and coal dust. Even the fertile minds of Homer and Hesiod couldn't have dreamed up such power as what Earth's tectonic forces are capable of accomplishing with ease. The New Madrid Fault had erupted, thrusting up a bulwark to dam the Mississippi and reverse its course for a few hours, giving proof that history always trumps mythology.

The February 7 quake was only the last in a series of monumental explosions that wracked the American heartland. It all began at around 2:00 a.m. on December 16, 1811 with a sudden release of seismic energy that sent out shockwaves strong enough to whip strong trees back and forth throughout southeastern Missouri and northeastern Arkansas. Six hours later, with barely enough time for the sparse population then residing in those territories to have caught its breath, another equally energetic onslaught transpired. After a respite of several weeks, the seismic assault continued on January 3, 1812 with yet another colossal earthquake. All four of these quakes were world-class contenders. No one knows exactly how strong they were—many seismologists are content to leave them somewhere between magnitude 8.0 and 9.0. All four of them warped the ground, created fissures, caved in river and stream banks, sent up unstoppable jets of ejecta, saturated the air with such a blanket of noxious vapors as to cause darkness, and more or less pummeled the area with such ferocity as to knock books, plates, and glasses off shelves on the East Coast and to crack sidewalks almost nine hundred miles away in Washington, DC. The shock emanating from the epicenters ranged at least 1,500 miles. The shaking was strong enough to set church bells ringing in Boston and Toronto, topple chimneys in Maine, and awaken sleeping residents as far away as Pittsburgh and Norfolk.

The New Madrid Fault Zone is enigmatic. One glance at the map in chapter six tells why. There is no plate boundary running through the mid-section of the North American continent, and common sense dictates that such a prerequisite is necessary in order to produce an earthquake. It seems as anomalous as witnessing a hurricane being whipped up on the Great Lakes. Yet earthquakes do indeed take place in the heartland of the United States, and these temblors can obviously be in the same league with some of the strongest shakers that have ever occurred in any part of the world. How that is possible is still a matter of scientific deliberation.

In chapter one mention was made of "cratons." The term was coined by geologist Leopold Kober in the 1920s. It means "strength" in Greek. When Earth's crust first formed there were no continents, just the newly formed basaltic crust which was constantly extruded and subducted again by the conveyor-belt dynamics of tectonics. As lighter, more silicon-rich granitic minerals formed, islands of this "froth" rose to the surface and resisted being pulled back down into the bowels of Earth, insuring their existence by their relative buoyancy. These ancient accretions of lighter materials that simply refused to be pulled back down into the mantle from which they were born became the seeds of the platform for life on Earth. There are three dozen or so cratons. They are the bedrock of the continents.

Over hundreds of millions of years they have joined together, split apart, rejoined, and split again to form geography in the past that our eyes will never be able to map with surety. Science does, however, have some idea of what the face of Earth's surface may have looked like going back at least 750 million years and here is where attention should be drawn regarding the New Madrid Fault Zone.

What is now North America used to be part of the supercontinent Rodinia. Seven hundred and fifty million years ago, Rodinia began to break apart. The surface under what is now Missouri, Arkansas, and Mississippi, in response to such enormous pressure, also began to rift. By the luck of the draw, it didn't have quite the impetus though to pull it off. Two hundred million years ago, now as part of Alfred Wegener's supercontinent Pangaea, another stab at secession was ventured, and also failed. What remains of these foiled attempts to break apart is what geologists call an "aulacogen," a primeval scar, a fractured weak zone in the bedrock, a demarcation line of something that might have been but ended up stillborn. This rift would be considered relatively seismically quiescent, yet the vestiges of the plate boundary that never was—despite trying twice to go its own way—can be reactivated from time to time, brought back to life by the east-west compressive forces the continent undergoes as it drifts. The North American plate isn't sitting calmly undisturbed. It is engaged in a constant shoving match with adjoining plates. These immense shocks and tensions cause compressions that must radiate the length and breadth of the continent. The only place for it to buckle is in its weakest region, its Achilles' heel, its hidden aulacogen beneath the Mississippi Valley. And the same primal pressures from deep in the mantle that have instigated and abetted two previous efforts to part ways shouldn't be forgotten. There's as much reason as not to assume that a third is underway at this very moment. Hence, this area, known as the Reelfoot Rift, is under siege from all sides *and* from below. It's little wonder that it can and does produce "intraplate" earthquakes—and that some of them should be veritable monsters.

Populations in the Mississippi Valley—in Memphis, St. Louis, and dozens of other cities—are likely less interested in primordial geologic history than in how future events might unfold. Certainly, there will be more earthquakes in the central United States, big and small, so it is important that people in this area become aware of the danger and how to minimize it. Cities in California have evolved with special construction codes, have passed retrofitting laws, support companies who offer foundation and home inspection services, and have inculcated their citizens with the need for emergency preparedness, all in response to having put down roots in an extremely active seismic zone. The West Coast is exponentially better prepared for whatever the San Andreas or Cascadia might throw at it. San Diego might shrug off the same sort of jolt that would conceivably cause severe damage in Evansville, Indiana or Jonesboro, Arkansas. Nature, too, has stacked the deck negatively. Earth's crust in the eastern United States is much older, colder, and denser than the crust in the West. Seismic waves traveling through this medium are stronger and range much further.

There is one more point of great concern for central North America. As the Soviets discovered and as so many dam construction projects have proven, injecting water under pressure into the underground environment of a seismically active area is tantamount to pulling on a lion's tail. That, however, is just what is transpiring at this moment in the New Madrid Fault Zone.

Hydraulic fracking involves drilling wells deep into the ground (seven or eight thousand feet, or even more) and then injecting liquids under extreme pressure in order to extract hydrocarbon fuels. Petroleum, natural gas, and other valuable resources can be acquired in this way. There are a number of techniques in use, but something common to all of them is a fracturing of bedrock and a breaking down of layers of sediment beneath the surface—all accomplished by the use of fluids under pressure. This is by no means a new technology, but in very recent years there has been an explosive revival of fracking due to the rising costs of fuels. There are fields being exploited now all over the United States, including Arkansas, Missouri, Mississippi, and California. This is currently a hot-button topic and many argue that the benefits are outweighed by the health risks: tainted ground water, toxic fumes, etc. France, Bulgaria, and some other countries have actually banned it. Reporters have written stories in Pennsylvania about people turning on their faucets and having flames shoot out.

In addition to reports of health concerns, there has been a veritable explosion of reputable media stories on the seismic dangers of fracking. Henry Fountain of the *New York Times* is one of hundreds of examples. He wrote "Expert Says Quakes Tied to Gas Extraction" (October 22, 2011). The expert was Brian Baptie, a project team leader with the British Geological Survey. The American response also was unusually forthcoming. Steven Horton, a seismologist with the University of Memphis, is quoted as saying, "The conclusions are reasonable." Dr. Horton investigated a swarm of earthquakes in 2010 and 2011, including one of magnitude 4.7, in an area of central Arkansas where fracking was in full swing. His verdict was that the earthquakes were probably caused by the disposal of waste liquids from the process. *USA Today*, *Huffington Post*, CNN, Reuters, *Discovery*, *Los Angeles Times*, *Scientific American*, and hundreds of others have published similar pieces. The best evidence concerning the nexus between fracking and earthquakes is found in legislation enacted in New York. The Empire State banned fracking in the summer of 2015, publicly announcing seismic connections as one of the primary concerns for the moratorium.

This, however, is in no way a misguided clarion call to pull down our dams or ban fracking; nothing of the sort is intended. Even if the human race were somehow determined to lurch back into pre-electric times and forego the epoch-changing benefits of hydrocarbon fuels, all for the quite chancy benefit of lessening potential seismic danger, the miniscule benefit would be overwhelmed by a tsunami of suicidal liabilities, for we would have to do without much more than gasoline and hydroelectric power. Many think that the greatest natural disaster in Australian history, the 1989 Newcastle earthquake, was precipitated by human engineering expertise, specifically by coal mining activities near Boolaroo, New South Wales, the epicenter of the earthquake. So going down this road—no dams, no mines, no electricity, no fuels, no raw materials—leads nowhere. The point to be made, however, is not a political one aimed at any particular industry or infrastructure. Almost every planet-wide undertaking by mankind comes with attendant risks, a subject far removed from this work. But, certainly, the outdated view that there's no way to predict an earthquake is turned on its head, since the declaration would be made by one of the great, sure-fire agents to cause earthquakes: the human race. And it's not just earthquakes that humanity now has the power to precipitate—not at all.

Earth's ozone layer is a roughly twenty-five-mile-thick section of O_3 (three atoms of oxygen per molecule) located in the tenuous environs of the upper atmosphere. It became thick enough to start shielding land-dwelling life on the surface from harmful doses of ultraviolet rays from the Sun about 700 million years ago, and still does. Chlorofluorocarbons are, on the other hand, quite a recent chemical formulation, first produced in 1928 for the air conditioning and aerosol industries and only vented into the atmosphere in sizeable quantities starting in the 1960s. Yet British scientists studying the ozone layer in 1985 discovered that this seemingly eternal feature of Earth, extant for the last half billion years, and which had taken perhaps two billion years to form, had already been partially destroyed in the space of two decades of chlorofluorocarbon use.[15] The original hole in the ozone layer, discovered over Antarctica, wasn't the only one, either. Damage to the ozone shield was later found over the Arctic and elsewhere.

Continent-sized tracts of old growth forests have disappeared in the path of human populations—and with startling rapidity. Western Europe's primary forestland, for example, intact in Roman times, basically vanished between the years 1100 and 1500. Poor farming practices in the 1930s contributed to turning 100 million acres of the heartland of the United States into a dust bowl, and then, after being remedied, just as quickly brought back to a man-made sea of grain. Today, over 150 million acres are given over to corn and wheat alone.

---

[15] Joe Farman, Brian Gardiner and Jonathan Shanklin, *Nature*, Volume 315, May 1985, pg. 207–210.

Human engineering has changed the physical map of the globe. The Salton Sea, Lake Meade, and countless other bodies of water are man-made. Conversely, the Aral Sea has been reduced to a fourth its original size by having its feeder waters diverted for agriculture. Artificial islands have been built in the Persian Gulf near Dubai. Well over half the population of Holland (60%) lives on land that used to be under the Atlantic Ocean, following centuries of dam building that pushed back the sea and increased the size of their country by 40%. By draining swamps, changing the shape of deltas, straightening rivers, building breakwaters, and countless other types of projects, over the centuries the human race has remade the very face of Earth in whatever image was required.

In fact, mankind's engineering has reached a level sufficient to actually change the velocity of rotation of the planet. Dr. Benjamin Fong Chao, a geophysicist at the Goddard Space Center (NASA), calculated that the ten billion metric tons of water (2.4 cubic miles) held in the dams and reservoirs built in the last century has acted to increase the velocity of Earth's rotation. As ice skater's spin is quickened as she draws her arms in closer to her torso, so does Earth react similarly by having so much water evaporated from equatorial regions and impounded in dams at higher latitudes. This redistribution of weight has shortened our day by two millionths of a second.[16] The extent to which dams control the movement of water is evidenced by the fact that 60% of Earth's river flow has been changed by humans.

The biology of Earth is also subject to the inadvertent and

---

[16] "Dams for Water Supply are Altering Earth's Orbit," *New York Times*, March 3, 1996.

deliberate whims of humanity. It has been hypothesized that mammoths, short-faced bears, giant ground sloth, and other mega-mammals disappeared because they were hunted to extinction by Neolithic tribes. Whether that's true or false, there isn't the slightest doubt about how the following group of animals met their collective end: moa, Tasmanian wolf, English wolf, quagga, Turanian tiger, Caspian tiger, Steller's sea cow, dodo, speckled cormorant, Pallas's cormorant, Irish deer, passenger pigeon, great auk, Caribbean monk seal. The Romans captured and slaughtered so many lions that the species was wiped out completely in North Africa. In one series of games alone in AD 107, the emperor Trajan caused more than eleven thousand animals to be slain to celebrate his victories in Dacia.[17] Today it is impossible to determine the number of threatened and endangered species, as the numbers have spun exponentially out of control. It is enough to say that there are many biologists who believe that we are currently living through one of the greatest episodes of mass extinction in the planet's history—this one caused by neither comet nor terrestrial disaster, but by human beings.

That mankind has the power to create or destroy islands and seas, punch holes in the stratosphere, shorten the day by making the planet spin faster, turn desert into farmland and forest into pasture, and wield the power of life or death over every creature that shares this planet suggests that there may be other unintended surprises in store. That human activity might have some connection with seismicity is hardly surprising.

---

[17] Shelby Brown, "Death as Decoration: Scenes of the Arena on Roman Domestic Mosaics," in *Pornography and Representation in Greece and Rome*, edited by Amy Richlin (New York: Oxford University Press, 1992), 4–73.

Northern Lights, viewed from the International Space Station (Courtesy of NASA).

Chapter Ten

Earthquake Lights

As earlier noted, many religions and mythologies say that the human race was first crafted from lowly mud. That dirt itself truly holds the wonder credited to it by sacred texts and legends. It contains the unheralded savior of all land-based life on Earth: silicon. It is silicon-rich granite, composed of 72% silica (silicon dioxide), whose buoyancy, compared to iron-heavy and magnesium-heavy basalt, allowed the continents to form.

Silicon has finally come into its own, but not owing to its service to life billions of years ago; silicon is a semi-conductor. Untold billions of diodes, transistors, and computer chips are fashioned from this nether material—neither conductor nor insulator of electricity, but a little of both and not very good at either. When semi-conductors are "doped" with small amounts of other elements—boron, gallium, phosphorus, or arsenic—an amazing thing happens. Due to a quirk of chemistry having to do with the four electrons in the outermost shell of silicon atoms, there will be either too many or too few electrons in that exterior bonding shell when traces of the doping elements are introduced, creating so-called "holes." Now current can flow, not just in one but in both directions, making silicon a semi-conductor. Currents running through silicon can be amplified, which allows it to be used to build transistors and solid state electronics. Silica-rich granite also naturally contains doping elements, some abundant and others in traces. So why isn't granite a transistor? Well, it may be, in a manner of speaking, under the right circumstances.

Dr. Friedemann Freund is a NASA principal investigator. For the last three decades he has been a lead scientist at the SETI Institute in Mountain View, California and an adjunct physics professor at nearby San Jose State University. Dr. Freund is a world-class expert on doping and electron holes, and with regard to the study of seismic forecasting, he has discovered a most unexpected dopant element: oxygen. Recently Dr. Freund placed a block of red granite under a 1,500-ton press, equal to the titanic stress experienced as tectonic plates twist and contort bedrock just prior to earthquakes, and discovered something rather remarkable. Voltage built up on the rock's surface, while sensitive cameras developed at Pasadena's Jet Propulsion Laboratory simultaneously recorded infrared emissions issuing from the hyper-strained rock. If current is somehow propagating through semiconductors below the surface of California and other seismically active regions, no one knows for certain exactly how the electricity and infrared light is produced, but Dr. Freund has a remarkable answer: water.

Subterranean rock soaks up vast amounts of water. Under the extreme heat and pressure at depths, the water molecules' bonds are broken, releasing the constituent parts, hydrogen and oxygen, and in the process setting into motion a complicated series of transformations. These traces of water form hydroxyls, essentially an oxygen with an imbedded proton making an OH–. A series of chemical changes then converts hydroxyl pairs into hydrogen atoms and peroxy bonds, and when the peroxy bonds are at last broken, the free electrons and "holes" are both available to do some fairly interesting things.

"It's similar to how an electrical charge radiates through a battery," says Freund. In my conversations with Professor Freund, he has taken great pains to explain this unusual type of battery:

> When rocks are stressed and the peroxy bonds sever, the electrons stay within the stressed rock volume, while the "holes" have the amazing ability to flow out into and through adjacent unstressed rocks, traveling fast and far, many feet in laboratory experiments, miles to tens of miles out in the field. In this way the stressed rock volume acts like a battery, which has a certain amount of electricity stored. In a regular battery the charges that flow out are electrons. In the stressed rock battery the charges that flow out are "holes." Because they have this remarkable property to flow out, they have been given their own name: "positive holes." As those positive holes [flow] through the Earth's crust, they constitute an electric current. All currents produce magnetic fields and any current that fluctuates emits electromagnetic waves. Hence, very low frequency electromagnetic waves in this fashion travel with little loss in their intensity through the rock column and finally arriving at the Earth's surface. If they flow into water, they turn it into hydrogen peroxide. When they arrive at the dry land surface,

> electrical potentials are built up, ionizing the air and discharging that ozone. The positive holes can also recombine, returning to the peroxy state, and in the process they emit infrared radiation.

Those processes creates a subtle, infrared glow very much like the one Dr. Freund has produced in the laboratory, and according to the premise, the emissions should be able to be detected one to two weeks before a major earthquake. The infrared light shines into space where Freund proposes to place an array of twenty dishwasher-sized satellites in order to capture the signature—an early warning seismic system, as it were. As for what his scientific peers think of the theory, Dr. Freund makes a salient observation: "Geology journals weren't interested because it was semiconductor physics, and physics journals weren't interested because it was geology." What truly sets Friedemann Freund apart from every other scientist is that he's willing to put his money where his mouth is—both literally and figuratively. Dr. Freund has donated more one million dollars from his own personal funds to kick-start the project and to convince NASA to take up the challenge. There is evidence that the money may be well spent. Electrical currents in rock could account for the numerous occasions in which magnetometers have recorded slight fluctuations in Earth's magnetic field prior to major earthquakes. The most famous incident concerned the Loma-Prieta earthquake that struck San Francisco in 1989. Low-frequency magnetic signals were picked up for two weeks prior to the quake, twenty times above normal levels, and then intensified even more on the actual day of the quake.

There is doubtlessly some phenomenon linking seismic activity with visual displays in the sky, so called "earthquake lights." There are far too many occurrences—witnessed by throngs of observers—of lights being splayed across the sky just prior or during great earthquakes. These incidents happened in the United States, China, Japan, Chile, Peru, Canada, Italy, Turkey, and elsewhere. In fact, the first mention of these lights is written in ancient Greek and is found among the list of seismic alert signals catalogued by Pausanias in the second century AD in his *Hellados Periegesis* ("Description of Greece," 7.24.7–8): "Warnings are wont to be sent before violent and far-reaching earthquakes . . . occasionally great flames dart up and across the sky." Dr. Freund's explanation for the lights seems reasonable enough: "When a powerful seismic wave runs through the ground it compresses the rocks with great pressure and speed, creating conditions under which large amounts of positive and negative electrical charges are generated. These charges can travel together, reaching what's called a plasma state, which can burst out and shoot up into the air."

The speculation, however, is that the source of these lights is far, far beneath the surface, accessed by extraordinarily deep faults, reaching down sixty or seventy miles, the lights the result of tiny defects in the crystal lattice of deep basaltic rocks, which give off brief electrical charges when stressed mightily. "We speculate that the charges are focused until they become an ionized solid-state plasma," says Dr. Robert Theriault, a geologist with the Quebec Ministry of Natural Resources in Canada. "When the plasma bursts into the air it produces light."

In any discussion of light—infrared, the glow from ionized plasma, or any other kind—it may be well to cast an eye on the greatest source of blistering electromagnetic energy in this small corner of the universe: the Sun. There are a number of researchers who have hypothesized that solar activity can and does precipitate earthquakes on Earth, via solar flares and coronal mass ejections. Whether that's true or false, what is certain about the Sun is that nothing need be invented about its influence on Earth in almost every realm; the plain facts are quite enough.

Solar disturbances send out powerful magnetic waves and greatly increased solar winds. In the 1800s solar storms regularly knocked out telegraphic communication in all parts of the globe. During World War I, Germany's intelligence-gathering capability was hampered when the country encountered solar-generated radio interference so severe that the nation was virtually blacked out for a period in 1915. In 1921 a terrific storm burned out a great portion of the signaling and switching system of the New York Central Railway, set fires in control stations, and wrecked telephone, telegraph, and cable infrastructure all over Europe. The auroras produced by the "Fatima Storm" of 1938 were so intense that people all over Europe assumed their cities were ablaze and sent fire brigades dashing in every direction. Firefighters in London were dispatched to Windsor Castle, determined to protect this English icon from the non-existent blaze. The 1940 Easter Sunday Storm burned out generators, fused cables, silenced the US Coast Guard's radio stations, and hushed almost two hundred thousand miles of the Associated Press's land lines. The solar storm that Quebec experienced in 1989 caused the power grid to collapse entirely, casting six million people into the dark in the middle of winter. So what about seismic activity? How could the Sun cause an earthquake ninety-three million miles away from its surface?

Geomagnetic storms on Earth occur when the solar wind pressure is increased in response to great outbursts on the Sun, effectively compressing Earth's magnetosphere. The shock wave driving everything can come from coronal mass ejections (CME), increased stream of solar wind due to flares, or enhanced densities in the solar wind formed at the interface of a faster solar wind stream overtaking a preceding slower wind stream, a phenomenon known as co-rotating interaction regions (CIR). Milder interactions cause stronger auroras over the poles; stronger storms can melt transformers, as noted above. Even more powerful storms may drive down into the center of our planet—the spinning iron core, the ferociously hot source of electromagnetism on Earth—and cause other interactions. The core not only produces our magnetosphere, it is the engine that drives plate tectonics. It is hypothesized that great torrents of energy bursting into Earth's core may result in surges in the tectonic currents that eventually evidence themselves at the surface—as earthquakes and volcanoes.

These ideas, certainly controversial, have nonetheless been at least partially endorsed as plausible by some highly respected scientists at National Oceanic and Atmospheric Administration (NOAA), the US Naval Observatory, the Royal Observatory, and the European Space Agency. However, on the other side, there is the USGS's opinion, duly researched, that there is nothing to be seen here that might lead to any sort of earthquake prediction paradigm. Dr. Jeffrey Love and Dr. Jeremy Thomas led a team of scientists who compared historical data of solar activity with earthquake occurrences around the world and found very little to suggest that one directly influenced the other. Their findings were published in the *Geophysical Research Letters* in 2013. "Recently there's been a lot of interest in this subject from the popular press, probably because of a couple of larger and very devastating earthquakes. This motivated us to investigate for ourselves whether or not it was true," said Dr. Love. "It's natural for scientists to want to see relationships between things. Of course, that doesn't mean that a relationship actually exists."

Here is where the matter lies. It may be well to reiterate that in natural history on occasion reality can far outshine speculation. The great New Madrid earthquake of 1812, for example, not only caused the Mississippi River to run backward, but also created the world's largest sand boil, a slurry a staggering 136 acres in extent, of sand and water bubbling to the surface through liquefaction. It produced thousands of seismic "tar balls," golf ball-sized solidified petroleum nodules left strewn throughout boils and fissures. It let loose an "earthquake smog" thick enough to turn the skies so dark and black that at noon even lamps didn't help. Yet before the gloom descended and while the air was still transparent, many eyewitnesses reported the dazzling lights, across great arcs of the horizon, reaching from the ground into the sky almost halfway to zenith.

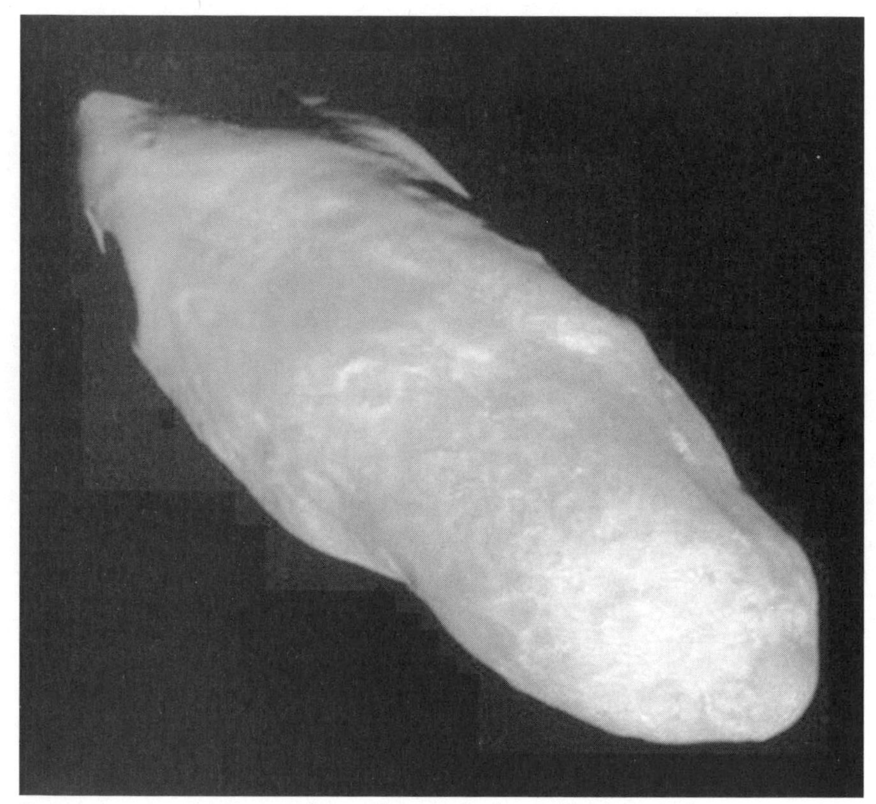

Saturnian moon, Prometheus, discovered in 1980 (NASA).

Chapter Eleven

4.5 Quintillion Joules

Joseph Stalin can be faulted for any number of terrible shortcomings. That's probably the kindest way to describe someone who sent some thirty million people to their graves. No one, however, has ever accused him of one of the worst deficiencies: stupidity. Stalin was taken aback that tens of thousands of his countrymen should have perished in earthquakes in the late 1940s and didn't waste too much time giving his ear to the excuses offered him by Soviet scientists as to why his government was blindsided by the events. Speaking as a simple son of a cobbler from Georgia, Stalin simply couldn't believe that monumental seismic eruptions could build up and explode without giving the slightest indication prior to the fact. It defied common sense.

Comrade Stalin may have had a point.

To pick up an apple resting on a table and raise it up to one's mouth so as to take a bite requires energy—not that much, about one joule roughly. Now, let's put Prometheus on that table. Prometheus is one of the moons of Saturn, named for the titan who turned his back on the gods in order to befriend mankind, stealing fire from the immortals to bestow on humanity. Zeus decreed he should suffer eternal castigation for the crime and chained him to a mountain in the Caucasus—according to the folklore of Stalin's homeland it was specifically the dormant volcano Mt. Kazbec, the third highest peak in Georgia. Every day an eagle arrived to feast upon his liver, and every night the organ grew back again so that the punishment should continue unending. Prometheus isn't a titan of a moon, but it's no chopped liver either—a heavenly body of more than a third of a million cubic kilometers. Imagine the muscle required to pick up a moon and yank it a full meter from the table to one's mouth in order to take a bite out of it. Nothing like that could happen on this planet? Well, it can and it has. Manhandling Prometheus like that would require either the strength of Zeus or 1.6 quintillion joules of energy—exactly the same calculus of energy expended in 1964 when the Great Alaska Earthquake set the record for the biggest recorded seismic event in the history of North America.

Such is the energy expended when Earth quakes. One must enter the realm of mythology and numbers with interminable strings of zeros attached and beyond to begin to comprehend the landscape. It is not a friendly place for the human mind. There are quite a few sharp scientific minds, however, that do in fact agree with Joseph Stalin that earthquakes must give warning signs. It does seem very much out of step with common sense that astounding telluric movements such as the Great Alaska Earthquake should transpire right out of the blue, with not a whisper in advance of their impending wrath, with no growl or grumble prior. It has been seen in prior chapters that there may in fact be precursors announcing the next seismic catastrophe. There are some scientists who believe Earth "hisses" before letting loose with its punches, so here is yet another theory to consider: geohydrochemical precursors.

Ernest Rutherford, the 1908 Nobel Prize winner in chemistry, has a number of accomplishments to his credit. He discovered the nucleus of the atom (Rutherford atomic model) and was probably the first person to split an atom in 1917. He is also the discoverer of radon. Prior to 1899 this colorless, inert gas was unknown. Radon is radioactive and it's also one of the heaviest gases—nine times denser than air. It exists in single-atom molecules, unlike H2 or O2, and so has the ability to penetrate materials easily—plastic, glass, concrete, wood, etc. Large reservoirs of radon are present in the rock and soil of Earth's crust, the gas itself the result of the natural radioactive decay of subterranean uranium and thorium. Radon seeps to the surface, can be dissolved in groundwater, and can also be introduced into the atmosphere via deep wells. This escaping radioactive gas is real enough. The Surgeon General of the United States estimated that currently close to twenty thousand deaths per year are caused by the carcinogenic effects of radon (Surgeon General's Advisory, January 13, 2005), representing $2 billion per year in health costs.

There are scientists who deem it feasible that cracking bedrock can announce an impending earthquake through the measurement of the fluctuations of radon gas emissions along a fault line. New fissures in splintering bedrock provide additional avenues for ground water to wash out greater surface areas of the substratum—a rock layer infused with radon. Such groundwater, when measured, should then show a spike in its radon content. But the back and forth since the early 1990s on the merits of radon as a reliable indicator are as argumentative as any in science, even after the L'Aquila earthquake. Christopher Scholz, a professor of Earth and Environmental Sciences at the Lamont-Doherty Observatory at Columbia University, along with two colleagues, was the first to propose that radon emissions might serve as earthquake warnings—forty-three years ago. When asked recently for his opinion of radon as a potential predictor, his response is that the theory still has yet to be proved or disproved. "There are so few measurements made that there's not a good database to say if it works or not."

It is estimated that nearly a third of the US population felt the magnitude 5.8 earthquake that shook a great portion of the US East Coast on August 23, 2011, more than any other earthquake in American history. (As mentioned in chapter 9, the older, colder, harder crust east of the Mississippi River greatly amplifies the strength and range of seismic waves as compared to California.) Chapman University geophysicist Dmitar Ouzounov and his team used the opportunity to look at seven years' worth of satellite images of the earthquake's epicenter, in the Piedmont region of Virginia about twenty miles southwest of Spotsylvania, and utilized a sophisticated computer program to estimate the average heat radiating at that spot each day. They found that thermal radiation spiked significantly for nearly two weeks before the quake, continuing right up until the earthquake itself. On the day of the tremor, they also saw electromagnetic changes near the site very high up in the atmosphere. They put forward the idea that heat spikes and electrical changes before earthquakes are caused by escaping radon gas that heats the lower portions of the atmosphere (although, as seen in the previous chapter, Dr. Freund might give a different answer for those same phenomena).

Radon isn't the only gas grabbing attention from researchers. Helium, hydrogen, carbon dioxide, and others are being studied too, along with mercury concentrations and other effusions in groundwater. The basic idea of geohydrochemical seismic precursors isn't that hard to grasp. If elements and compounds not normally seen to be venting or flowing to Earth's surface from deep within suddenly make their appearance, it's not that great a stretch to conceive that, using the words of Dr. Richards in speaking about the Deccan Traps, "something has changed the plumbing." Slippage, the opening of fissures, the fracturing of bedrock—some sort of movement or change must precede the uncorking of gases not seen on Monday yet pouring forth from the ground on Tuesday. And while researchers all have their favorite candidates for chemical precursors, there is one gas that seems to be drawing a consensus.

To say that helium-3 is rare is an understatement; it is astoundingly scarce. Until recently the only place thought to contain appreciable quantities was within the regolith of the surface of the Moon. The government of China is mounting a project within the next few years to establish a base on the Moon to harvest helium-3 and ship it back to Earth. Why would they contemplate such an intricate and expensive program? Helium-3 may turn out to be a key component in the next chapter of energy production: fusion reactors.

There is an international race being run among all the major powers to get the first fusion reactor up and running. A fusion reactor would leave our current fission models in the dust, not unlike comparing a uranium fission bomb to the mega-destructive power of thermonuclear fusion devices such as the fifty-megaton Tsar Bomba—four thousand times more powerful than the Little Boy bomb used over Hiroshima. Fission involves splitting an atom, while fusion requires that two atoms be squeezed into one. In effect, a fusion reactor would be a mini-sun built here on Earth, turning isotopes of hydrogen into helium, creating blistering heat in the process, using that heat to flash water into steam to spin turbines that in turn would churn out vast amounts of electricity. There are just the few small problems to solve in order to create a man-made star.

Atom smashers have recorded temperatures in the trillions of degrees—a quarter million times hotter than the center of the Sun—but only for infinitesimal durations and heating volumes far smaller than an atom, useless for producing a single watt of electricity. A tokamak is a machine built to really do the job and the progress being made here is simply astonishing. The record for the highest sustainable temperature created by human ingenuity at the moment is 510 million degrees centigrade. Tokamaks hold the broiling plasma they create in the only way possible—by not actually allowing it to touch anything. Magnetic confinement is utilized to keep materials twenty-five times hotter than the core of Sun within the "walls" of invisible magnetic fields. The hydrogen isotopes found in "heavy water"—deuterium and tritium—have been deemed the appropriate fuels for fusion. Once reactors are a reality, they will be as ubiquitous as the seawater from which heavy water will be extracted to power civilization. Helium-3 comes into the picture for a few good reasons. If helium-3 replaces tritium, its fusion with deuterium might be more efficient, it would generate no radiation, and it would produce no radioactive waste materials.

Even though it is a gift of the Sun, delivered within the solar wind, heliuim-3 is rare on Earth due to the planet's magnetic shield, which prevents it from being deposited on the surface. The

Moon, however, possessing no such defense, has been bombarded by angry alpha particles, including helium-3 nuclei stripped of their electrons, for more than four billion years. The Chinese think there's around 1.2 million tons of the stuff infused into the first few centimeters of the lunar surface. Other known locales for stores of helium-3 are even more difficult to access: the gas giants in the outer solar system—Jupiter, Saturn, Uranus, and Neptune. That all has changed though, and only recently was it discovered that helium-3 can be found somewhat closer to home: it's seeping out of the ground in Southern California.

In the summer of 2015 a University of California at Santa Barbara geologist, Jim Boles, made the startling discovery that helium-3 was leaking from deep oil wells along a thirty-mile stretch of the Newport-Inglewood fault zone in Southern California. There is only one source for helium-3 on Earth and it's unfortunately in a place more difficult to access than the lunar surface: the upper mantle. Coming not from the Sun nor any other star, the helium-3 inside our planet was born in the Big Bang itself, part of the protoplanetary disk of debris that made up Earth and the rest of the solar system. That the gas should be able to escape from such depths—at least twenty-two miles beneath ground—is proof positive that the Newport-Inglewood fault is a much, much deeper gash in the crust than had previously been thought.

What this means is that this ancillary fault of the San Andreas is capable of bigger earthquakes than seismologists had considered possible, and that is no unimportant piece of information for a fault that lies under the second largest city in the United States, one of the world-class centers of culture and commerce of the Western world: Los Angeles. What Newport-Inglewood has already wrought in the past should be considered sufficiently attention-getting. The epicenter of Long Beach earthquake of 1933 was on the southern segment of this fault; 120 people were killed in this second most deadly earthquake in California history.

A virtual ocean of petroleum has been pumped from beneath the surface of Southern California. Indeed, Los Angeles used to float on top of it, the gooey substance oozing from the ground in many places (like the famed La Brea Tar Pits). Knowledge of these seeps goes back to colonial times, the Spanish reporting how the native tribes used the pitch to waterproof their canoes. Aside from this use, the substance was regarded as more of a nuisance than anything else until 1892.

The story goes that the luckless prospector, Edward Doheny, was in downtown Los Angeles when he noticed a delivery cart with pitch smeared all over its wheels. The deliveryman pointed Doheny to an area near present-day Dodger Stadium, where the first successful well was dug, supposedly with a sharpened eucalyptus tree. Two years later eighty wells were producing crude oil, and by 1900 there were over 500. The oil boom was on.

By 1910, California was producing seventy-seven million barrels of oil per year. In the 1920s, one-fourth of the entire world's petroleum supply came from beneath California, the state pumping approximately 250 million barrels yearly. From the 1890s to the present, billions of barrels of petroleum have been extracted.

In decades to come, as the oil runs out and the world moves on, it may be something else extracted from beneath Greater Los Angeles that might fuel the next energy bonanza. Southern California's fault lines may pay for themselves after all, even deducting for their intermittent seismic temper tantrums. While the Chinese may have to travel all the way to the Moon for what may be fueling our world in the next century, Angelenos might be able to find theirs after nothing more than a short journey over to Culver City or Inglewood.

Half a world away from Los Angeles is the only place where it is possible for Earth to stage demonstrations to not only equal what happened in Alaska in 1964 but far outdo them. Chile's earthquakes definitely fit into the monster category. The one that occurred on May 22, 1960, magnitude 9.5, the largest shaker ever recorded anywhere, exerted enough energy not only to pick up Prometheus but to twirl it around and juggle it with ease: 4.5 quintillion joules. If one is interested in studying big earthquakes, Chile is the destination of choice. Something even more ethereal than gas seems to be bursting from epicenters of large Chilean earthquakes just prior to their rupturing (corroborated by other great tremors around the world), and the data is indisputable—detected by the American, European, and Russian space agencies and published in unshakably qualified publications. "High-energy particle bursts" are associated with some truly major-league seismic events. Here's what NASA has to say about the matter, quoting Arkadi Moiseevich Galper, Principal Investigator at Russia's National Research Nuclear University: "This study of seismic events shows that it may be possible to predict earthquakes based on bursts of high-energy charged particles in near-Earth space. Scientists mapped the location of particle bursts, then analyzed correlations between those and known quakes. They identified particle bursts that could have predicted earthquakes registering more than 4 on the Richter scale."[18] The

---

[18] "Monitoring of Seismic Effects-Bursts of High Energy Particles in Low Earth

following list of the earthquakes, which were preceded by sharp disturbances in the ionosphere and that were tracked back to their source on Earth, is one worthy of note: Alaska, 1964 (131 deaths); Indian Ocean earthquake and tsunami, 2004 (280,000); Haiti, 2010 (316,000); Chile, 2010 (550).

The first satellites to discover these disturbances linked to seismic activity were those of NOAA, in March of 1964, so this is not a new phenomenon. It is only the worldwide attention that is new. Even as far back as the mid-1980s, the Russians were aware of the connection. Concerning the MARIA experiment conducted aboard the SALYUT 7 orbital station, "The results of these experiments confirmed the correlation between short-term sharp increases of particle intensities and seismic processes. Moreover, it was found that the particle flux variations appeared twenty-four hours before the main shock of strong earthquakes. This means that the earthquake precursors in the near-Earth space were experimentally observed. This emission, as it was shown in some of the ground-based experimental observations, was created in the earthquake epicenter several hours before the main shock."[19]

DEMETER (Detection of Electro-Magnetic Emissions Transmitted from Earthquake Regions) is a French satellite placed into orbit in June of 2004. It was on duty on February 27, 2010 when the magnitude 8.8 blockbuster detonated just off the coast of Chile.

Space Region," International Space Station Program Science, Principal Investigator, Arkadi Moiseevich Galper, Nov. 22, 2016.

[19] "High-energy Charged Particle Flux Variations in Vicinity of Earth as Earthquake Precursors," A.M. Galper et al, Moscow State Engineering Physics Institute, January, 2001.

DEMETER recorded the energetic charged particle bursts over the epicenter of the seismic activity starting eleven days before the quake in the energy ranges between about four to six times average value.

It may well be that Earth has been speaking loudly and clearly, giving out plain warnings before letting loose, issued in gases and other emanations that spring from the very heart of the planet, and that until now few credible listeners could be found. That is far from the case now. Earth has an eager audience at last, hanging on her every word.

What is causing these bursts to take place in the high atmosphere, registering at satellites in orbit, is quite a story. Dr. Cristiano Fidani is a distinguished professor at the University of Perugia in Italy, attached to the Andrea Bina Seismic Observatory. He set me straight about the real source of these particle bursts. One shouldn't imagine that a fountain of energetic sub-atomic particles are streaming from openings deep within the bowels of Earth just prior to eruptions, making their way up and out through newly opened avenues to the surface and then escaping into space to be detected by satellites. If that were the case, he told me, "they would be absorbed by the first few centimeters of the atmosphere." What is being emanated is a bit more ethereal: ultra long frequency (ULF) radio waves. This is the primary source for the phenomenon, according to Dr. Fidani. The use of the term "ultra" is not hyperbole; it's hard to imagine light stretched out to such extremes. ULF waves range between 3 kilohertz (wavelengths that measure 100 kilometers from crest to crest) all the way to radically super-extenuated waves a thousand kilometers in length in the 300-hertz band, electromagnetic waves that measure the distance between New York and Chicago from crest to crest. Just a bit shorter are VLF waves (very long frequency), radiation squeezed just a bit tighter but still cutting quite a swath: from 100 kilometers to 10 kilometers. The ULF waves impinge on charged ions in the rarefied regions of the high atmosphere and it is this secondary reaction that gives rise to the bursts. The energetic particles aren't created below ground, only the mechanism.

All of this speculation about ULF waves is interesting enough in and of itself, but there is much more to it. These ridiculously long waves, created by dynamics that may take place just prior to earthquakes, have garnered the attention of some very powerful agencies, the United States Air Force among them. In chapter 14 we'll investigate how some governments on Earth are exploring how to potentially use them—not to forecast great eruptions on the verge of explosions, but instead to determine the possibilities of harnessing the vast powers of seismicity for their own purposes.

San Francisco, 1906 (Library of Congress).

Chapter Twelve

Saint Andrew, Rock Striker

Saint Andrew was one of the twelve apostles, the patron saint of Russia, Ukraine, Romania, and Scotland. According to legend, a ship transporting the saint ran aground in Cyprus. Andrew came ashore and struck the rocks upon which the ship crashed, and a spring of healing waters gushed forth. The captain of the errant ship who was blind in one eye made use of the pool and his sight was restored. In the New World, the most famous geologic fault line in the Americas, the San Andreas Fault, is named after this biblical rock striker.

Many dozens of great cities in California are at risk to some degree from ground-shaking movements along this eight-hundred-mile-long boundary between the North American and Pacific Plates. Quite naturally the subject of when the next "Big One" should strike is a source never far from the public forum or the minds of tens of millions in California. There is a great deal known about when and how and where the next great movement on the San Andreas should happen, and the unadorned facts are far more scintillating than anything Hollywood can dream up.

When the Next "Big One" Might Strike

In the late 1970s, Dr. Kerry Sieh and his colleagues dug trenches across Pallet Creek in California and took samples from ruptures in the layers of peat. The seismologists were able to prove that the disturbances in the peat were the vestiges of ancient earthquakes transpiring on the San Andreas Fault in the past. That the deeper breaks corresponded to the older earthquakes was apparent. Thanks to radio carbon-14 dating, however, a much better appreciation of the age of each offset in the peat layers (and hence the date of the earthquake that caused it) was established.[20]

Geologists Tom Fumal, Silvio Pezzopane, Ray Weldon, and David Schwartz dug similar trenches across a creek near Wrightwood, California more recently. They were able to determine, using the same methods as Dr. Sieh, that twelve "Big Ones" have occurred on the southern sections of the San Andreas Fault in the last 1,300 years. The results were published in *Science* in 1993, with the authors concluding in the final paragraph that "the data shows that for the last five earthquakes on the southern San Andreas fault northeast of Los Angeles the recurrence interval has averaged around 100 years, significantly shorter than the 132 years reported by (Dr. Kerry) Sieh, et al., and the elapsed time of 135 years since 1857."

The authors refer to 1857 as the year of the last "Big One," the

---

[20] Peat, after all, is organic material and a perfectly acceptable substance for carbon-14 dating. While an organism is alive, the level of carbon-14 in its tissues remains constant. Upon death, carbon-14 diminishes by virtue of radioactive decay. The half-life of the unstable isotope of carbon is roughly 5,750 years. Without going into the mathematics involved, the less carbon-14 remains in a sample being studied, the greater its age.

Fort Tejon earthquake, which was 135 years before their writing in the early 1990s. So, proximate estimates of periodicities on the San Andreas have at least been sketched and published in peer-reviewed scientific journals, giving an inexact answer as to when great ruptures might take place. In 2017 it will be 160 since the last great rupture on the southern San Andreas. The very simplest calculus as to when the next should occur is uncomplicated: it should have already transpired. Southern California is overdue for the next great telluric explosion.

Determining the locations more prone for such occurrences would be as important as determining the timing, and seismologists have been pointing for a long time at the exact stretches of the fault line where a great quake might more likely to occur. Unfortunately for the population of Southern California, the most worrisome lengths of the San Andreas are the very segments closest to one of the most densely populated areas in the United States.

Where the Next "Big One" Might Strike

To determine where the next great tremor in Southern California will occur, seismologists have been using Seismic Gap theory. The idea is that if a certain segment of a known fault is observed to have remained quiescent over a long period of time, it is reasonable to assume that portion of the fault line to be locked tight. As time goes on the tension only increases in these locked segments, as do the probabilities for an eventual explosion, since the seismic energy simply continues to build rather than being released in small and frequent jolts, for centuries in many instances, until a catastrophic rupture occurs. Data regarding the normal rate of movement between the plates in question, the size of the earthquakes usually produced, and the history of seismic activity all along the length of the entire fault line gives seismologists a way to roughly forecast where the next temblor might occur. A group of scientists did just that with regard to the San Andreas Fault in the 1980s. The Working Group on California Earthquake Probabilities studied the lengths of the fault and published probability estimates connected with the likelihood of a large earthquake being produced in each segment. The map that follows shows where the Working Group established the highest hazard warnings (the Mojave and Coachella Valley Segments), in close proximity to Greater Los Angeles, the second most populated area in the United States.

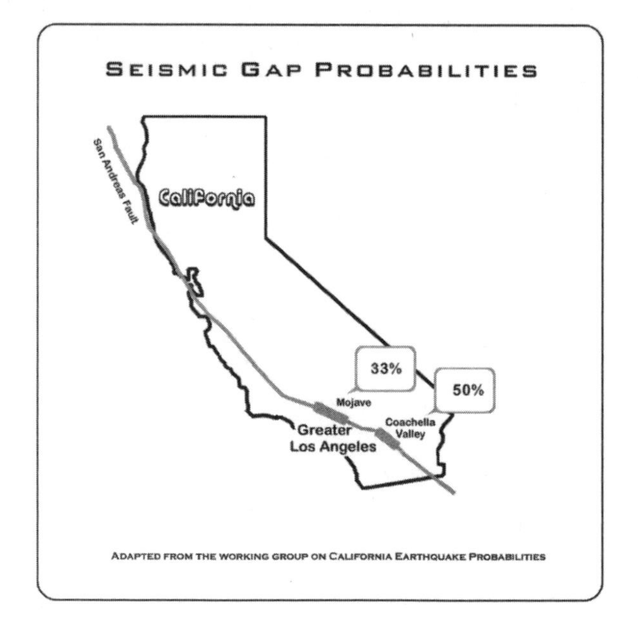

Author's collection.

The figures speak as starkly as the map: a 33% chance for seismicity on the Mojave Segment between 1988 and 2018, and an almost 50% chance for the same on the Coachella Valley Segment.[21] (Only the very short Parkfield Segment had a higher hazard rating.) The sections of the San Andreas in question, abutting Greater Los Angeles and San Diego, have been loading more and more seismic energy, century after century, and still have not ruptured. USGS says they are "primed to break." Indeed, that is an apt description. It is estimated that the last seismic relief along some stretches of the fault line was around 1690. As far as Southern Californians are concerned, and especially those residents nearest to the Coachella Valley segment, it might be well to put into historical perspective how long it has been since that section has sundered. Peter the Great was czar of all the Russias the last time it fractured; some estimates actually date the last event on this segment of the San Andreas at a time just before Columbus landed.

How the Next "Big One" Might Strike

The Fort Tejon earthquake of 1857 was a magnitude 7.9 quake—and it was just a regular, run-of-the-mill "Big One." It was powerful enough to beat the tops of trees against the ground for three minutes. Lake Tulare was shaken so violently that its waters were splashed from its shores for up to three miles in every direction. The next day people gathered fish that had been stranded in the

---

[21] Working Group on California Earthquake Probabilities, "Seismic Hazards in Southern California: Probable Earthquakes, 1994-2024," *Bulletin of the Seismological Society of America*, 1995, pg. 286–291.

upheaval by the wagon-load.[22] One eyewitness to the event, J. M. Barker, a young ranch hand that had gone riding early that very morning to the shores of Lake Tulare, searching for missing cattle stock, commented later that "We could only imagine what the consequences would have been if a great city had stood upon the shores of the lake."[23] Great cities do stand near there now: Fresno to the north and Bakersfield to the south—and an even greater city further south still: Los Angeles.

It's been calculated by the USGS—based on the San Andreas's length, depth, and class (strike-slip)—that earthquakes larger than magnitude 8.3 are extremely unlikely. Since the Richter scale is logarithmic, Californians should certainly hope the next great rupture on the San Andreas is commonplace rather than record setting; an 8.3 earthquake is four times more energetic than one in the 7.9 range.

According to USGS's figures,[24] a 7.8 earthquake on the San Andreas near Greater Los Angeles will shake the entire region for a mind-numbing duration of three full minutes. Some freeways will come down; some high-rise buildings will collapse. Injuries could be in the range of 50,000 casualties, absolutely swamping hospitals throughout the Southland, and with 1,800 fatalities. About a thousand of those deaths will result not from the earthquake itself, but from the 1,600 fires that will rage out of control, immediately stretching Southern California's firefighting apparatus to the

---

[22] Eyewitness account of J. M. Barker, recounted by Donald Eisman, "Fort Tejon Earthquake of 1857," *California Geology*, August 1972, Volume 25, No. 8.

[23] Ibid.

[24] Associated Press, "Scientists Detail Impact of Big One," May 21, 2008.

breaking point and far beyond. This, it should be stressed, is not the "worst case scenario," USGS cautions, but only one with a "plausible narrative that would have major consequences."

How the Empire is Striking Back

As worrying as the foregoing may seem, this is hardly a one-sided battle, and in truth the San Andreas isn't facing the same landscape it squared off against in the nineteenth century. Los Angeles will be armed with some rather amazing defenses this time around. For starters, there is no city better prepared to take an extreme seismic punch and somehow still stay on its feet. Decades of preparations in stringent city building codes, the undertaking of mammoth retrofitting programs, and nonstop preparation and practice in every segment of public and private life has created a city that may be able to withstand almost anything the San Andreas can dole out. Just as one might expect from the classic thrillers that have been created in Los Angeles for cinema for decades, just at the right moment, striding onto the stage just when California needs them the most, comes a group of truly heroic saviors.

The United States Geological Survey has most certainly come to the rescue. If I have disagreed at times with USGS, it is easy to make this simple and honest appraisal: USGS, at end, is a living, breathing example of the best of the United States of America. Judged against the panoply of often meddling, useless, and burdensome bureaus thrust upon the backs of Americans, USGS—along with NASA, the Army Corp of Engineers, and a few others—can arguably be counted among the greatest creations of our republic. Since 1879, this agency has been the piston in the engine that has developed our energy, discovered our resources, and managed our wildlife. Their motto is "Science for a changing world," and the world *has* changed. USGS has created a noteworthy array of instrumentation that could be considered the foundation for an early warning system. An impressive network of creepmeters, tiltmeters, strainmeters, gravimeters, electronic distance meters, and other sensitive instrumentation has been put in place such that Earth can scarcely wriggle without monitoring stations all over the world knowing about it instantly. There were a few hundred seismograph stations in existence in the United States in 1931; today there are many thousands, and they are interconnected with USGS's partners around the world to tens of thousands more internationally. Such a network, already up and running, makes USGS aware of the slightest changes in the status quo beneath the crust of Earth along the San Andreas, the Cascadia, and New Madrid Fault Zones and elsewhere. Deformation monitoring sites can detect minuscule variations in topography, perceive every new bulge even if such protrusions be measured in the millimeters, determine the torque and pressure changes on massive slabs of bedrock deep beneath the ground—and even evaluate how Earth's gravity and magnetic field might be altered in localized areas due to the unseen yet perceived rumblings in the bowels of Earth.

The sincerest accolade of all, however, concerns the stellar achievement USGS has only recently accomplished: full-fledged, dead-certain, indisputable, completely reliable "earthquake prediction." USGS has achieved this feat by "cheating" a little, but against such an opponent as the San Andreas the gloves are definitely off. An Earthquake Early Warning system is nearing completion, currently stalled by money issues. This network will detect the P-wave signature of large earthquakes at monitoring stations in remote areas all along the San Andreas and Cascadia Fault Zones as they happen and then send electronic warnings to race ahead of the seismic waves. (Along the West Coast the network has some 650 stations in place; it's estimated that another 1,000 are needed to fill in quite sizable gaps.) Since those alerts will travel at light speed while the energy of the quake itself moves at the speed of a satellite in orbit, cities in the path of destruction could receive precious time to take last-minute cover—even a few seconds could make a difference. The further out from the epicenter, the longer the lead time for the public to take cover. Schools, businesses, mass transit, airports, hospitals, first responders, utility companies, and many millions of individuals connected to the system, whose cell phones will issue alerts seconds before the arrival of the seismic waves, will be given at least a short time to take actions that may spell the difference between an unmitigated catastrophe and a far less damaging disaster. Doctors only seconds away from performing intricate surgery would be able to halt, as well as pilots preparing to take off or subway conductors leaving stations. In the near future it's quite possible that gas companies may install automatic valves to respond to these signals by shutting off supplies. Historically, it is almost always the case that terrible damage and dramatically increased death tolls from great earthquakes are the direct result of fires that burn out of control in cities that have lost their power to fight them due to the shattering of water mains. Turning off the gas—and snuffing out tens of thousands of potential fires before broken gas lines can be ignited—would alone be a staggering game-changer in the history of earthquake preparedness.

Granted, this system's effectiveness will be greatly diminished concerning the earthquakes that strike directly below Greater Los Angeles, the Bay Area, or Seattle/Tacoma. Even if the system had been in place for the entire duration of the twentieth century, no warnings could have been issued in time to make any difference at all for Northridge, Sylmar, or any great tremor whose epicenter was under cities to start. One can't send a warning about an earthquake on its way to the city under which it is transpiring. Nonetheless, Los Angeles, San Francisco, Seattle, and other great cities on America's West Coast will soon be joining other world-class cities—Mexico City, Istanbul, Tokyo, etc.—with this twenty-first-century seismic protection.

The San Andreas is almost never the source of seismic assaults in California, though. There are more than 15,000 ancillary fault lines that spread out from the main fault line—the San Andreas—under California alone. It is upon this spiderweb of secondary faults that almost all earthquakes strike, and new ones are being discovered all the time. Northridge struck on a fault that was unknown until Martin Luther King Day of 1994. Many of them are big enough to cause great damage, in the magnitude 6.0 to 7.0 range. Northridge, Sylmar, Long Beach, Coalinga, Whittier Narrows, and other killer quakes give evidence of that.

Chinese researchers and others have noted changes in the electrical conductivity along faults just before earthquakes. USGS is aware of this phenomenon, but the mechanisms at play are not understood as yet. Water, once again, may be a tool to determine what could be transpiring beneath these fault lines prior to great seismic events. As seismic strain increases on a fault under extreme stress, the intense heat and pressure beneath the surface can force water held in rocks in deeper substrata to be released and concentrated in cooler rocks that overlay closer to the surface. Scientists who measure the electrical conductivity of fault lines may see a change in the resistance as they pass current along a certain segment now holding more water. Or it may be that the resistivity of the rock itself is changed by virtue of the awesome compressive forces. The University of California at Riverside has installed a telluric current-monitoring array in Parkfield, California and is currently attempting to solve this mystery. Dr. Steve Park, then a geophysicist at UC Riverside and currently at the Massachusetts Institute of Technology, told the *Los Angeles Times*, "We are looking for changes in the Earth's electrical resistivity, or how well electric current flows through the ground. Natural currents are flowing because of electromagnetic waves in the Earth. Think of rock as a sponge, filled with water. Electricity is conducted within a fluid. It is more difficult in a solid. If you squeeze the sponge a little, you change the pore space and you change resistivity."

However, no brain trust at the University of California or MIT, no power on Earth or elsewhere is going to impede the unstoppable progress of two incomprehensibly massive plates—the Pacific and North American—as they slide under the cities of tens of millions of people on the US West Coast. And yet, while the San Andreas may have won the first few rounds, the human race has faced down far greater foes than the San Andreas, and been bested by none of them. The bubonic plague depopulated whole continents in its time, yet is now on the verge of extinction, hiding from our antibiotics, infecting the odd rodent here and there on the Eurasian steppes. It's much the same story for smallpox, polio, and a long list of former terrors that have long since been forced to submit to a clan of gutsy anthropoids who might get knocked down from time to time, but who always struggle back to their feet, swinging hard enough to defeat any opponent foolish enough to climb into the ring, surrendering to the implacable courage and intelligence of our forebears.

The greatest weapon any enemy possesses is that of surprise. It's not 1857 anymore. So when the San Andreas or some ancillary fault radiates its next burst of city-shattering power into some sleeping community on the West Coast in the quiet hours around dawn, thanks to mankind's vastly improved knowledge regarding tidal triggers, geohyrochemical precursors, electromagnetic indicators, its array of satellites and ground-indicating sensors of every kind, it may be that some of the citizenry will be found awake, prepared—alert.

Mt. Saint Helens, 1980 (USGS).

Chapter Thirteen

Mother of the Orphan Tsunami

I have been privileged to maintain a correspondence with Dr. Vinayak Kolvankar for many years. He is the retired Chief Seismologist of the Seismology Division of the Bhabha Atomic Research Center in Mumbai, India and the author of quite a few scientific papers published in the *Journal of Indian Geophysics Union*, *New Concepts in Global Tectonics Newsletter*, and other peer-reviewed journals. A few years ago one of Diane Sawyer's producers at ABC World News asked me to relay an invitation to Dr. Kolvankar to consider going on record with that respected news organization concerning his views on tidal interactions and earthquakes. He didn't actually say "no," but I could easily tell by the tenor of his response that he wasn't very interested and the matter was dropped. When I informed him that this book was in the works he sent me a note reiterating his opinion. "During our studies we've noticed that Sun-Earth-Moon positions and distance feature a relationship with earthquakes. It is astonishing to see that worldwide earthquakes faithfully follow this straight-line relationship. This proves beyond doubt that a vast majority of earthquakes are governed by the Sun and Moon. Even the smaller earthquakes in the magnitude range of 2.0–3.0 follow this relationship."

When Dr. Kolvankar refers to "Sun-Earth-Moon positions," he's pointing to new and full moon phases. With regard to variations in distance, however, a bit of explaining is required. The Moon isn't always the same distance from Earth. The Moon doesn't orbit Earth at the same distance throughout the month. It veers sometimes closer and sometimes further away from the planet in its circuit. Perigee is the term for its closest pass to Earth; apogee is when it is most distant. The tidal influence, however, is not measured at one-to-one correspondence. The tidal effect of the Moon's perigee is exponential, increasing or decreasing at a rate of the inverse cube of the distance. That is to say that if the Moon were to circle Earth at half the distance it currently orbits, its tidal effect wouldn't be twice as strong, it would be eight times as strong (2 × 2 × 2). The Moon doesn't swing that erratically toward Earth as it orbits, but it does dip from around 406,500 kilometers to roughly 356,600 kilometers, substantially increasing the tidal stresses with its proximity. The Moon's tidal interaction therefore is increased by 37% when it is 10% closer at perigee as compared with its distance and tidal strength at apogee. These perigee events obviously happen once a month. From time to time they take place simultaneously with a new or a full moon phase, heightening tidal pressures to the maximum. Much more rarely, two such combined new or full moon phases in perigee± will occur in two consecutive months.

On February 17, 1980 at 8:46 UTC, the Moon was in perigee, and also entered its new moon phase twelve hours thirty-four minutes later. On its next circuit, March 16, 1980 at 20:30 UTC, the Moon was in perigee again, and was anomalously also only ninety-eight minutes away from precise new moon phase as well. Forty-three hours and fifteen minutes later, on March 18 at 3:45 p.m. (PST), a magnitude 4.2 earthquake centered below Mt. Saint Helens's northern flank marked the end of almost a century and a quarter of dormancy and was the incipient event for two months of almost non-stop tremors and venting, culminating in the deadliest and most economically destructive volcanic eruption in US history.[25]

Considering volcanic explosions in Washington brings us to focus on an area probably facing some of the greatest seismic risk on Earth—while at the same time, strangely enough, hardly on the radar screen of the American public's attention. A massive adjunct of the Pacific Plate, the Juan de Fuca plate, is being thrust beneath North America in the areas around Portland, Seattle, and Vancouver, British Columbia—subducted into the bowels of Earth, much like Lucretius had hypothesized two thousand years ago (the Cascadia Thrust Zone). It causes the coast of the Pacific Northwest to spring back every now and again, after fruitless attempts to drag the entire coast with it on its path to oblivion into the sea of magma beneath the West Coast. Only a thrust fault like the Cascadia can produce mega-quakes that can range into the magnitude 9.0 to 9.5

---

25 "Mt. St. Helens: Facts about Deadliest US Volcanic Event," *USA Today*, May 15, 2015

categories, as the Cascadia is quite capable of accomplishing. It has in the past and it will do so again.

We know the last time this occurred was on January 26, 1700, at around 9:00 p.m., from the historic evidence of Japanese villages on the other side of the Pacific that were destroyed by the "Orphan Tsunami" caused by this event. Chronicles in Japan note the previously puzzling history of a tsunami that struck seemingly out of nowhere around midnight on January 27, 1700. Ten-foot-high waves flooded rice paddies, washed away buildings, and knocked over lanterns that set fire to twenty houses in one town alone. Villagers, merchants, samurai, and others, primarily in the Japanese prefecture of Iwate, recorded the flooding and damage while taking note that the parent earthquake that should have given rise to the tsunami had not been felt by anyone, causing questions as to what had precipitated the waves. The tsunami did have a "mother," but this one was on the other side of the Pacific Ocean, in North America's Pacific Northwest. Brian Atwater, Kenji Satake, and others put the clues together and solved the three-hundred-year-old mystery in 2005 in their book, *The Orphan Tsunami of 1700*. By working backward from when the waves struck Japan and calculating the speed at which tsunami waves traversed the Pacific Ocean, they were able to nail down almost precisely when the great Cascadia earthquake had transpired. Studies of tree rings in Oregon and Washington, along with sediment layers, core samples from the ocean floor offshore, and debris samples from earthquake-induced landslides all dovetail (approximately, of course) with this time and date. Moreover, oral traditions of the peoples living in the area when the earthquake struck also support January 26, 1700 as the date of the massive tremor. Almost all the tribes living in the region have oral histories recounting a great earthquake around this time, far more powerful than any other, and some are buttressed further by temporal clues indicating how many generations ago the event took place. On Vancouver Island, Kwakiutl and Cowichan stories, and from Washington, Makah and Quileute legends all tell of a great winter quake, striking in the nighttime just as people had gone to sleep—causing landslides that buried whole villages. The Huu-ay-aht recount that every single person living along Pachena Bay on Vancouver Island perished; all drowned in the tsunami save for the one village of Masit, thanks to its location some seventy-five feet above sea level. A woman by thc name of Anacla aq sop, by fortune away when the disaster occurred, garners a poignant place in the oral traditions as the only surviving member of her community. From a tribe living on the inner coastline of Vancouver Island come these words passed down from three centuries ago: "The people could neither stand nor sit for the extreme motion of the earth."

It was certainly one of the most powerful earthquakes ever produced on Earth, in any epoch, in any locale—between magnitude 8.7 and 9.2. There are more than a few seismologists who believe that such a recurrence might not be too far over the horizon (previous earthquakes are estimated to have occurred in AD 1310, AD 810, AD 400, 170 BC and 600 BC), and unfortunately, many engineers and safety experts think that the Pacific Northwest is hardly prepared for such a cataclysmic event. What is doubly troubling is that an ancillary fault of the Cascadia Thrust Zone, the Seattle fault, runs directly beneath some of the most populated real estate in the Pacific Northwest. A great event on the Cascadia fault, or even a lesser one on the Seattle fault, would rival anything ever seen in California.

In 2008, a consortium of USGS seismologists, engineers, and safety experts (Scenario Group), after three years of studying the problem, released their best estimates as to what might be wrought by a 6.7 quake on the Seattle fault. It was published in the *Seattle Times*, although it could have been used as a screenplay for the next blockbuster mega-disaster film. In short, Seattle would be left in ruins. Seattle has 800 to 1,000 old brick and masonry buildings built in the 1930s; many of them would come tumbling down. Jing Yang at Cal Tech pointed out in her doctoral thesis that of the 900 high-rise buildings in Portland, Seattle, and Vancouver, many were constructed with brittle welds, and many of these would more than likely fail too. Sixteen hundred people might be killed, 24,000 injured, 45,000 families left homeless, and as many as 200,000 buildings damaged to some degree. Ferry terminals could slip into Puget Sound; freeway bridges and the Alaska Way Viaduct might collapse. There's the danger of large sections of Harbor Island slipping into Elliot Bay, taking the Port of Seattle's infrastructure with it. The Olympic Pipeline, carrying gasoline and jet fuel, could very well rupture, being laid across the mushy and unstable soil of the Renton and Kent valleys. Mark Stewart, of Washington's State Emergency Management Division sums up his opinion of the likely scenario in two words: "Pretty ugly."

Further south, the city of Portland can't be considered immune from the same danger. Yumei Wang, of Oregon's Department of Geology and Mineral Industries told the Associated Press (March 2, 2010) that close to 1,300 schools and public buildings in that state would be at risk in the event of a sizeable seismic event. Oregon's recorded history documents fairly large temblors—a 5.9 quake in 1993, a 6.8 quake offshore in 1910—and there are three faults that run under Portland itself. Portland's soil, however, is loose sediment deposited during the last Ice Age, a factor in causing damage from even moderate shaking.

Science has only recently come to posit that some indications of how a great earthquake in the Pacific Northwest might be preceded by "silent earthquakes" taking place far below at the edge of the Juan de Fuca Plate as it is being dragged into the magma of Earth below the North American Plate. "Slow slip" earthquakes occur at certain parts of the two plates' boundaries that get "stuck," and these slow-motion earthquakes can last from hours to months as the seismic struggle plays out deep below the surface, some fifteen to fifty kilometers below ground. Again, it is well and good that these unusual tremors proceed at a snail's pace, slowly and safely expending all the pent-up energy over long durations. If the unhurried and steady rupturing of the encumbrances hanging up the Juan de Fuca and the North American Plates were to fail, the result could only be something other than slow: instantaneous. When the North American Plate snaps back with a vengeance, when the uncoupling between the two massive slabs of Earth resembles a jerk rather than a protracted grinding effort, that is when the greatest earthquake will happen: a megathrust earthquake. There is debate on all this, of course, concerning whether slow-slip events increase or decrease stress in complex fault systems, and even more dispute about how to determine where the most probable epicenter for a megathrust quake might be established by monitoring such arcane movements taking place some 150,000 feet below the surface—a depth sufficient to accommodate 120 Empire State Buildings set one upon the other.

What is certain, however, is that the Cascadia *will* erupt again. Since this thousand–kilometer-long gash in Earth runs between forty and eighty miles offshore of the Pacific Northwest coastline, it has the length, depth, and megathrust muscle to precipitate a devastating double punch: astoundingly powerful earthquakes coupled with mind-boggling tsunamis (the 1700 tsunami reached estimated heights of thirty-three feet along the Washington coastline).

USGS estimates that the probability of that happening within the next fifty years at 40%.

Aftermath of tsunami, Great Alaskan Earthquake, 1964 (USGS).

Nikola Tesla (Courtesy of the Library of Congress).

Chapter Fourteen

Loose Lips

Nikola Tesla was certainly one of the great geniuses of the modern age. It's not hyperbole to opine that our world would be a different place without him. A partial list of his accomplishments is enough to make any normal person blush for shame in comparing his or her average mundane life. His preeminent invention, the electric dynamo, is enough to put him in humanity's hall of fame. He also invented the neon lamp, wireless telegraphy, and of course the Tesla coil, along with obtaining a staggering three hundred patents in his life. The entire megalithic superstructure of anything and everything having to do with alternating current can be traced back to the foundation laid by this one man.

One of those inventions was the electro-mechanical oscillator, or earthquake machine. To understand how Tesla attempted to create a temblor, imagine a father pushing his young daughter on a swing. In order to keep the girl happy and provide a thrilling ride, the man can't just thrust his arms out any time at all. He'll know to wait until the girl first reaches the height of her outward progress, comes to a momentary stop, is pulled back toward him by gravity, and comes to another stop, hanging motionless in the air in front of him. Then will be the time to invest another burst of energy to set another cycle in motion. He'll want to keep his pushes "in phase" with the swing's oscillations. Waving his arms around when the seat of the swing is nowhere to be found is useless. Worse, pushing her when she's still falling back to him will "dampen" the cycle, and probably cause the little girl to frown. Another classic example of the same sort of intensified force by stages has the opera singer hitting just the right pitch, sending out sonic waves at just the right frequency to interact with the natural resonant frequency of the glass, causing it to vibrate at greater amplitudes, finally shattering it. The Tacoma Narrows Bridge was destroyed in just the same way in 1940 by the phenomenon of aero elastic flutter, in wind only blowing at forty-two miles per hour yet giving repeated nudges in just the right way to build to the breaking point. Tesla's idea was that what could be achieved with glass or steel bridge girders could be accomplished with any material—including bedrock. He tested his machine in New York in 1898. What happened next may be urban myth or it may be fact. The story goes that first his building, then others nearby, began to shake violently enough that police were summoned. The dénouement of the tale is a perfect fit too. The only way to turn the machine off was with—a sledgehammer.

Whether fact or fancy, what isn't up for debate is what happened almost a century later in 1996, which gives some insight into at least one plausible reason for reticence on the part of the United States government to give plain, simple, and straighforward answers to questions about the possibilities concerning earthquake prediction. In that year a group of US Air Force officers (three colonels and four majors) presented to the US Air Force a study entitled

"Weather as a Force Multiplier: Owning the Weather in 2025." Excerpts from the disclaimer on the title page state that the study was designed "to comply with a directive from the chief of staff of the Air Force to examine the concepts, capabilities and technologies the United States will require in order to remain the dominant air and space force in the future." To be clear, the views expressed are the authors' and "do not reflect the official policy of the Air Force, Department of Defense, or the US government." It's also perfectly acceptable for this matter to be addressed in the open as the publication had been reviewed by security authorities and "is unclassified and is cleared for public release." An overview of what the report contains can be given by simply quoting a few of the chapter titles: Messing with the Weather; Weather Modification; Artificial Weather; Precipitation, Fog, Storms: Why the Ionosphere is Important.

Now this is all very new, and all very old. Any military historian can give hundreds of examples of battles won and lost due in part to some clever commander's use of the weather conditions. Polished shields and armor and battle lines arrayed to take advantage of the rising sun to blind or confuse an enemy, naval engagements offered or refused while waiting for advantageous tides, night assaults scheduled during the darkness of a new moon, and countless other stratagems have been used over the centuries. The Battle of the Bulge, to give one of the best examples, was launched on the Allies when the German High Command had been assured by meteorologists that an extended period of "Hitler weather" was at hand (snow and overcast skies) so as to negate the overwhelming American and British air superiority. So if the Air Force is looking into the future and preparing to deal with the next generation of sneaky tricks, this should come as no great surprise. If the US Air Force isn't, there's a good chance that some other nations' armed forces are.

Cuba in the 1960s and 1970s might already have been on the receiving end of some of those schemes. Such mainstream news sources as *Newsday* and the *San Francisco Chronicle* (January 10, 1977) ran stories that lend credence to rumors about US involvement in the terrible swine flu outbreak in Cuba in 1971 that resulted in the loss of half a million pigs, supposedly introduced by clandestine airborne operations. In June 1976 United Press International printed an accusation that the United States seeded clouds in the 1960s and 1970s in order to squeeze rain out of them before they arrived over the island so as to damage the sugar cane harvests. The same article avers that cloud seeding over the Ho Chi Minh Trail muddied the enemy's main transportation artery.

Since all is fair in war, and if those who wish peace had best prepare for it, what is being considered is the possibility of man-made storms, droughts . . . and earthquakes. Over the years quite a number of missives have come to my attention via contact links at www.earthquakepredictors.com from people all over the world wishing to share information on this subject. Many of the writers have presented quite tolerable cases for their point of view regarding this matter. In 2012, I was invited by one of the UK's iconic radio presenters, Peter Price, "Mr. Liverpool," to share some of his air time. The topic of the show was supposed to be seismic forecasting on the West Coast, but when I parenthetically made mention of some of the "tips" to which I'd been apprised, Mr. Price was flabbergasted and immediately asked if it would be all right to change the tenor of the discussion. "The idea of weather warfare; I find that absolutely astounding!" he said. The rest of the interview was devoted to informing an English audience of what American readers should find equally interesting.

So just how could weather warfare plausibly be accomplished? Tens of thousands of conspiracy theorists have a simple answer: HAARP. This US Air Force experiment (High Frequency Active Auroral Research Program), from 1993 to 2014, tucked into the hinterland of Alaska, was an array of hundreds of antennas, each capable of putting out sizeable watts of electromagnetic energy, yet when focused and functioning as a single generator, the output went from sizeable to rather stupendous. (The numbers are left unspecified as the real figures are likely classified, known to but a handful of humans, and would probably not match anything anyone is likely to read here or anywhere else.) Extremely high frequency waves were beamed at the ionosphere at outrageously high power levels (over a billion watts, at least?), heating that area of the upper atmosphere so vigorously that it was converted to the plasma state. As the target area heated, it lifted, warped, and bulges creating a temporary "mirror" resulted. The Air Force says these reflectors, Artificial Ionospheric Mirrors (AIMs), could be useful from the standpoint of both bouncing communications to submarines using very long frequency (VLF) radio waves around the world and disrupting enemy transmissions, and leaves it at that. These are the same VLF waves that were discussed in chapter 12, the close cousin of electromagnetic radiation that satellites are observing being emanated by seismic areas just prior to earthquakes and giving rise to high-energy particle bursts in the ionosphere: ULF waves. At the

30-kilohertz wavelength boundary, for example, there's really no way to distinguish one from the other. Regarding the Air Force's supposition that submarines under many meters of sea water are still capable of receiving messages sent in that medium, there is quite a bit of evidence. VLF waves can certainly penetrate water. Bedrock doesn't hold them back either.

The evidence for VLF waves penetrating deep within Earth is close to a century old and is established by experiments conducted from far and wide. Those experiments were certainly not intended for weapons use. In fact, it was due to mankind's altruistic nature that the quest to determine the penetrating power was first set in motion. In 1930, Wallace Joyce, a Bureau of Mines geophysicist, conducted tests with VLF radio transmissions to ascertain whether they could be used to communicate with trapped miners in the event of cave-ins. Even with the paltry and primitive apparatuses at his disposal almost nine decades ago, he was able to pierce to the depth of a quarter of a kilometer with them. The South Africans did far better in 1949, with T. L. Wadley going nine times deeper, all the way down to over two kilometers. What the current record is in the early twenty-first century is unknown.

Apart from using AIMs for enhanced submarine communication, more suspicious voices point out that by uplifting the upper layers of the atmosphere, an area of low pressure is created by pulling up air from below. Clever manipulation of low-pressure cells could result in diverting water-laden jet streams to cause droughts or to nudge or strengthen hurricanes that have already formed to crash into specified landfalls. Katrina, for example, while wandering around the Gulf of Mexico, would have lost her impetus had it not been for an unfortunate fact. She passed over not one, but two ephemeral "hot spots" that are scattered around the Gulf. A hurricane running out of thermal energy can gas up at these places, drawing on warm water not only at the surface, but from the reservoirs that anomalously continue to the depths below. Katrina filled up her tanks, twice, and then made straight for New Orleans. No one has ever blamed HAARP for this or any other national or foreign agency. Katrina is nonetheless an excellent illustration of how nudging hurricanes one way or the other could have an enormous impact on a nation victimized in this way, should it be possible to guide storms by human means.

Regarding the production of earthquakes, one more twist need be added. Once plasma reflectors have been created in the high atmosphere—of just the right shape and size and height—the high frequency waves are turned off and the VLF waves are brought to bear. These very low frequency waves, the very band of electromagnetic energy produced naturally by Earth preceding earthquakes, could be generated to match the natural resonance of bedrock, and just like the father swinging his daughter, they bounce off the "mirror" in the sky and return to Earth to hammer away at a designated area on the surface, setting up harmonics that sooner or later might bring about a nasty seismic result. In fairness to the authors of the report, the foregoing description of how high-altitude mirrors might be created is not the method proposed in the prospectus. These artificial ionospheric mirrors are instead imagined to form by crisscrossing microwave beams that would ionize patches of the upper atmosphere. However created, either way, the job would be the same: to bounce VLF waves. The tenor of the report is left obviously nebulous (if the meteorological pun can be forgiven). For example, the idea of how storms might be whipped up or enhanced might "involve technologies to increase atmospheric instability" and additional "conditions" (the report itself has quotation marks around that word, whatever that implies). As far as how to steer a storm once it has been strengthened, that small problem is left open. It is indicated, however, that it would depend on which way the wind blew, which "currently is not subject to human control." The authors make it clear that they are looking thirty years into the future and admit at a number of junctures that they themselves have no clear idea where what is being proposed might lead, while intimating at the same time that to allow other potential unfriendly powers to take the lead in this field would be bad counsel indeed.

Archimedes said he could move the world given a lever long enough. Tesla said, "A ton of dynamite exploded at intervals of one hour and forty-nine minutes would amplify the Earth's natural standing wave until the planet split." Archimedes has been proven right; not many sane people would want to call Tesla's bluff, if indeed, that's what it is.

On August 1, 2012, the magazine *Despierten* published an article I wrote in Spanish for their readership in Caracas, Venezuela, "Terremotos, Predicciones y Possibilidades" ("Earthquakes, Predictions and Possibilities"). A few months later the editors sent me what to their minds seemed jarring indications that the United States had already begun engaging in seismic warfare. The data showed a list of earthquakes, all transpiring beneath the territories of certain nations that any unbiased observer would naturally describe as being inimical to the interests of the United States. The anomaly they wanted me to address was how it could be that so many earthquakes, striking in widely separated regions, could all have emanated from the exact same depth beneath the ground: at precisely the ten-kilometer level. This seemed to them the obvious fingerprints of human engineering. The official data checked out; all the temblors were in fact logged in databases as having transpired at the exact depth *Despierten's* editors claimed. Here, though, is a perfect example of finding what one is looking for, something politicians and quite a few others in the United States, Venezuela, and elsewhere (myself *not* excluded) are apt to engage in rather than following the problem solving dictum devised by William of Ockham, an English Franciscan friar of the fourteenth century. "Ockham's razor" is a philosophical principle that says that when faced with a battery of choices as possible solutions to some as yet unsolved problem, it is most likely that the simple explanation is the correct one. It turned out that the solution to this quandary was mundane in the extreme. From USGS's "Earthquake FAQs": "Sometimes data are too poor to compute a reliable depth for an earthquake. In such a case, the depth is assigned to be 10 kilometers. In many areas around the world, reliable depths tend to average 10 km or close to it. Thus, if we don't know the depth, 10 km is a reasonable depth. We used to use 33 km. Increased understanding indicates that 10 km is more likely."[26] Case closed.

It doesn't really matter at what stage the United States or our future enemies are with regard to weaponizing VLF waves, if that's possible. In 1996 the US Air Force at least started thinking about it, with the authors of "Owning the Weather" pointing up at the ionosphere and urging putting it to military use by 2025. HAARP then almost simultaneously went online in an inaccessible region

---

26 USGS FAQ: "Why Do So Many Earthquakes Occur at a Depth of 10 KM?" http://www.usgs.gov/faq/categories/9826/3427 February 3, 2016.

of Alaska. It's not unreasonable to wonder if the United States is attempting to discover a way to unleash weather war. Moreover, it is supreme artlessness to assume that our adversaries might not be exploring the same path. A weapon such as that could terrorize tens of millions of people living along the San Andreas, Cascadia, and New Madrid Fault Zones—not to mention the havoc it might bring along the Gulf and Atlantic coasts. Fortunately—from a military point of view—all the powers with the potential to develop their own VLF technology also possess their own vulnerable seismic zones, and their own susceptible coast lines. Mutually assured destruction between the Soviet Union and the United States worked to preserve the peace for decades in the last century. The threat of tit for tat, quake for quake, hurricane for cyclone might turn out to be part of the calculus of the near future. So if HAARP wasn't busily putting forth the effort to insure that the United States "remains the dominant force in the future," maybe they should have been. There's good reason to imagine that could very well be the case; there's very little the United States is willing to dismiss out of hand when it comes to national security. In fact, whatever else be said about the US military, it never met a potential weapons or intelligence gathering technique—either real or imagined—without immediately wanting to know everything about it. We can be thankful for such inquisitiveness, or else stealth bombers, patriot missiles, and Athena laser cannons wouldn't exist.

The United States government is both the most open-minded and closed-mouthed entity concerning the most seemingly bizarre and ostensibly left-field potential "superweapons." During the 1950s the United States was under the impression that communist regimes in Russia, China, and North Korea were using

LSD to brainwash captured Americans. Rather than fall behind in this new technique the CIA conducted Project MKUltra, experimenting with giving doses of LSD to unsuspecting subjects. In an even less flattering episode, from 1946 to 1948 the United States surreptitiously and purposefully delivered doses of syphilis to some seven hundred soldiers, prostitutes, prisoners, and mental patients in Guatemala and then attempted to cure them—killing eighty-three people in the process. In the mid-1940s penicillin had just been made available to the general public, and the government was extremely interested in whether penicillin could be used not only to cure but to prevent infection, what dosages of penicillin actually cured infection, and the process of re-infection after cures.

The Stargate Project was the code name for a US Army unit established in 1978 at Fort Meade to investigate the potential for psychic phenomena in military and domestic spying applications. This apparently harebrained project involved "remote viewing," the purported ability to psychically "see" events, sites, or information from a great distance. Stargate ran all the way to 1995 until a CIA report determined it was a completely useless endeavor.

As far as tectonic weapons are concerned, the United States was already seriously experimenting with attempting to create tsunamis—as far back as seven decades ago, during World War II and directed against Japan. Project Seal was a secret military operation conducted jointly by the United States and New Zealand. Tests conducted between 1944 and 1945 indicated that a line of four and a half million pounds of explosives detonated about five miles off the coast of Japan should initiate a destructive wave. Project Seal, a deadly serious attempt to unleash a manmade tsunami to wreak havoc along the Japanese coastline, was considered potentially almost as important as the Manhattan Project and only called off when events at Hiroshima and Nagasaki put an end to the war.

The American government alone shouldn't bear the entire onus for these and other missteps; there isn't a country on Earth whose secret files, if exposed to plain view, would elicit less opprobrium. It's the business of the powers that be of every nation to investigate any and all potential advantage against possible rivals and adversaries and over the span of decades or centuries there are few crimes and misdemeanors that aren't committed toward that end. But in this age-old, do-or-die affair of statecraft, the first order of business of any government is to say nothing. "Loose lips sink ships" applies not only to naval operations but is a dictum that can scuttle entire administrations as well. When our government or any other is forced to give answers, the responses are almost always pronounced in such a way as to confuse, befuddle, misdirect, and confound. So to ask the federal government of the United States for an open, complete, and unguarded appraisal of what it knows about earthquake prediction would be a fool's errand. When, where, how, and if the West Coast might be devastated by a powerful earthquake is a matter of the highest national security. Any government functionary who might honestly answer that question would find his or her career at an end, and for good reason. Aside from national security, there are other very grave matters at stake. The United States requires dams and other massive water projects whether they produce tremors or not. Our nation can't afford to cavalierly ignore the astounding quantities of natural gas and petroleum that fracking provides in order to prevent windows from rattling in drilling areas. The question of how mining operations may impinge on seismicity is not one that many governments care to debate.

These issues are not invited onto the public forum as matters for discussion. It can be no other way, nor should it. But to imagine this silence implies that the United States has no opinion on whether earthquakes can be predicted—or caused—simply shrugging its shoulders as if with no clue, is ludicrous. Even more preposterous is that the US government is lackadaisical regarding the safety of some seventy million Americans living in our seismic zones, a potential target for inimical powers that may now or in the future have the ability to cause earthquakes. This is the same government that chased illusory UFOs from coast to coast over two decades (Project Blue Book), that killed dozens of innocent Guatemalans in order to understand the finer points about penicillin, that sent unsuspecting victims leaping to their deaths from tall buildings in LSD-induced episodes of panic so as to potentially improve their spying techniques.

The crux of this matter, however, has nothing to do with droughts, hurricanes, weather warfare, or the machinations of global superpowers. Nor are the misdeeds of past regimes either foreign or domestic germane. It is simply that predicting earthquakes may finally come to pass. Massive reservoirs, deep drilling, hydraulic fracturing—and perhaps man himself, purposefully—might well be able to trigger earthquakes. Earthquake prediction as an impossibility might seem laughable in the future to military leaders who may be able to call one up at the commander-in-chief's orders.

Wright Brothers' first flight, Kitty Hawk, North Carolina, 1903, photographed by John T. Daniels. (Public domain image, via Wikipedia Commons.)

Chapter Fifteen

Scientific Method

The iconic photo to the left wasn't snapped by a *New York Times* photographer. It is one of the most dumbfounding incidents in the history of both photography and journalism that the man who took the image, one John T. Daniels, not only wasn't a professional but had never even *seen* a camera before in his life. He was merely someone who had accompanied the Wright Brothers to Kill Devil Hills near Kitty Hawk, North Carolina on December 17, 1903, and was simply dragooned for the job at the last moment. Not only was every reporter of the press absent, but neither was a single member of the scientific community present. The only five witnesses were Mr. Daniels and two others from the US government lifesaving crew, local businessman W. C. Brinkley, and a teenaged boy, Johnny Moore, who happened to live in the area. Even when the photograph was made available to the press and was sent out on the wires, fully half of the dailies in the United States refused to run the image, so certain were they that "if God had wanted man to fly He'd have given him wings." They preferred to believe the skeptical dictum rather than to trust their own eyes. In the previous chapter reasons have been suggested as to why the government should have little to say about seismic forecasting. There is likewise rationale for the scientific community's tepid and tight-lipped reaction to this discipline as well. They stretch far, far back into history.

On June 5, 1783, the Montgolfier brothers, Joseph and Etienne, simple paper makers by trade, sent a pig, a rooster, and a duck flying into the air as the first passengers in the world's first hot air balloon. News of this amazing event spread all over Europe within weeks and other experimenters accomplished similar results. A few months later the time had come for the Montgolfiers to attempt to send up the first human passengers. King Louis XVI was leaning toward the idea of using convicts to take on this dangerous task but was convinced by the Marquis de Arlandes that it would be a disgrace for France to bestow such glory on criminals. The Marquis himself, along with a member of the Academy of Sciences, Francois Pilatre de Rozier, were the first humans to ascend into the sky, on November 21, 1783, traveling five miles at an altitude of three thousand feet. The throngs of astounded Frenchmen witnessed on that day an event that had only previously existed in dreams and mythology, dreams old enough to go all the way back to images of humans with wings painted in Neolithic caves.

Benjamin Franklin was definitely present in the crowd that day, representing the United States of America as its ambassador to France. When a short-sighted acquaintance of Franklin's turned to him and asked of what use these balloons were anyway, according to legend, Franklin is supposed to have replied, "Of what use is a newborn baby?"

The genie *was* out of the bottle: it was possible to fly, after all. A great age of ballooning was opened by the Montgolfier brothers, using hot air and later hydrogen and helium as lifting agents. Balloons were used for military purposes, as no more excellent observation platform for directing movements of troops over large battlefields could be imagined. They helped to more efficiently increase the slaughter, especially during the American Civil War. By the 1850s the first steam-powered airships had been invented, and by the turn of the twentieth century the first years of the "Golden Age" of German zeppelins was inaugurated.

The pessimists of that age still insisted that this was "cheating," as it were. Men might *float* upon the air but they'd never be able to fly *through* it in a heavier-than-air machine. Leonardo da Vinci's schematics for just such a contraption—from as far back as the Renaissance—served to support their argument. Such baroque designs bore as much resemblance to reality as the gargoyles that decorated buildings from the same era. God neglecting to provide mankind with wings was their unbending motto. That it was bent and broken by two lowly bicycle mechanics from Ohio in 1903 at Kitty Hawk is one of history's great lessons, but still one that is probably not yet fully learned.

The human race's skepticism at its own prowess, the reluctance to rely on anything but recognized authority to paint visions of the future, the stagnation that comes from clinging to the known rather than exploring further has held back mankind more than any Dark Age could. There are many historians who believe that the Industrial Revolution could have—should have—begun two thousand years earlier than it did. From as early as the Hellenistic age, around 200 BC, humanity had all the tools necessary to jump-start a mechanized society. For example, Heron of Alexandria built the world's first steam engine seventeen centuries before

James Watt was even born. The mathematical prowess necessary to propel society from a classical to a modern age might well have existed also. Archimedes, the greatest mathematician and scientist of his day, was flirting ever so closely with the invention of calculus, and recalculated pi to its fifth correct decimal place (3.14159) using a method so similar to integration that it causes one to marvel that two millennia elapsed before Isaac Newton and Gottfried Leibnitz (1646–1716) finally came to the idea.

As much as modern culture shares with the ancients, there are still some places where the divergence is so profound as to stretch credulity. One such place concerns the chasm between classes. Artisans, craftsmen, and mechanics in the ancient world, even though they shouldered the burden of civilization, were looked upon with extreme scorn by the upper classes. There was an impermeable membrane that separated the purely intellectual realm of the patrician classes from the soiled, sweat-stained world of the plebeians. Cato the Elder, the quintessential Roman noble, could write treatises on every minute detail of how to run the successful, profit-making latifundia[27]—where to place the manure pits, how close the beehives should be to the farmhouse, what kind of trellis to use for the grape vines—with himself probably very rarely putting a shovel to dirt in his life. Bricklayers, carpenters, masons, glass-blowers, plumbers, potters, and blacksmiths were regarded as little more than the scum of the earth, yet the manuals these workmen followed were written by the same people who

---

27 Hooper, William Davis and Ash, Harrison Boyd. *Marcus Porcius Cato, On agriculture; Marcus Terentius Varro, On agriculture.* Volume 283 of Loeb classical library. Loeb classical library. Latin authors. Harvard University Press, 1934.

held them in such low esteem. This strange dichotomy brought about unhealthy results.

Many historians have connected the stagnation and eventual fall of the Roman Empire with this stark class divide. There is no use for a steam engine or any device like it when labor is performed by slaves. Heron's invention was looked upon as an interesting toy, something to mildly amuse. In fact, any labor-saving innovation would have been as welcome as an eye-poking machine, especially since the aristocrats had every reason to dread the idea of the millions of slaves in their midst being given some idle time. The Third Servile War (73 BC–71 BC), led by the gladiator Spartacus, was no mere slave revolt. It required the entire energy of the Roman army, led by the foremost generals of the day—Crassus, Lucullus, and Pompey the Great, to finally put it down.

From the beginnings of Western Civilization, where there was a noble elegance attached to theory, thought, and intellectual creativity, there was just the same opprobrium connected to anything that might come of it that required hammers or nails to prove. Aristotle got away with what seems now to be the ridiculous assumption that heavier things fall faster because not a single aristocrat to whom he was speaking had the slightest inclination to go outside and actually check it. Had Aristotle even deigned to address them, the craftsmen who might have checked it were too busy building the world. So while it was the great genius Daniel Bernoulli in the early 1700s who developed the mathematics to measure the lift exerted by an airfoil, it still was left to two grease-monkeys from Dayton to actually put it to use and build the first airplane. And although it was a French marquis and an academician who first flew, it was two provincial middle-class paper makers who built the airship they rode in. Such is the history of aviation—and many, many other sciences.

Any encyclopedia's entry on Michael Faraday will include the long list of accomplishments of this English scientist: in physics he discovered electromagnetic induction, diamagnetism, the laws of electrolysis, and invented the electric motor and generator. In chemistry he discovered benzene and two new chlorides of carbon and invented the Bunsen burner before Herr Bunsen did. Michael Faraday didn't even graduate the British equivalent of high school, and was ignorant of trigonometry and calculus. However, no example of this type, among the thousands that exist, can top the case of the Swiss patent clerk (third class) who was passed over for promotion to the exalted status of patent clerk, second class, in 1904. The very next year, 1905, Albert Einstein's *annus mirabilis*, he went on to other things, such as changing everything the world has ever known about light, gravity, time, space, energy, and mass. (Incidentally, in 1906 he finally *did* get the promotion to patent clerk, second class.)

The points to be made are threefold. Illustrious professors aren't responsible for every single brick ever laid in the scientific foundation of our society. The idea that only pure thought is worthy enough to contain truth is as bankrupt as the fraudulent notion that the handyman's crafty use of trial and error must surely lead to falsehood. And, more than anything, bets placed against the human race to overcome any obstacle in its path is money most assuredly lost. Yet the wagers keep coming.

The naysayers didn't lose a single beat after the Wright Brothers took to the air. Maybe it was possible to fly after all, but only so fast. There was an unbreakable sound "barrier" that couldn't be surpassed—that was nonetheless broken by Chuck Yeager even before the turn of phrase had time to pass into common usage in the English language. But, absent a sonic obstruction, humans could still not make it into space with a deadly Van Allen Radiation Belt forever hemming us in and preventing our escape from Earth. Somehow hundreds of American, Russian, and allied astronauts have managed to overcome even this. For the pessimists, though, our luck might just have run out—again. For, there's no possible way to conquer the stars; that much must *certainly* be true. The distances involved are mind-boggling, the speed of light impossible to reach, the prowess of our engineering unequal to the task.

A realistic appraisal says that not only is the colonization of a decent sector of the Milky Way by the year 3000 plausible, the better adjective is "likely." At this early date there already are very cogent plans to forecast just how that will be accomplished. *The Millennial Project,* by Marshall T. Savage, is a very convincing blueprint (among many others) that details a step-by-stepexpansion of humanity into the very quadrant of space that it is said to be impossible to go.[28] Incidentally, *The Millennial Project* is introduced by Arthur C. Clarke, who in commenting about the feasibility of Savage's vision simply says that he "wishes [he] had written it." Clarke could be considered a good judge of practicalities; he's the man most responsible for putting the United States' first communication satellite, Echo, in orbit.

It's been over four centuries since Giordano Bruno was burned at the stake in 1600 for declaring and refusing to recant

---

28 Marshall T. Savage, *The Millennial Project*, Little Brown and Company, New York, 1992

that Earth orbits the Sun. Since that time science has supposedly abjured doctrinaire dogma and has enshrined something to take its place: the scientific method. It is not inquisitors, or professors, or doctors of theology and science who now demand to explain how the world works and insist that their word be treated as holy writ, it is something far more concrete and yet sublime: hypothesis, prediction, observation, and analysis, the crux of the scientific method. If hypotheses concerning seismic forecasting are deemed absurd on their very face, if earthquake prediction is branded as impossible, a taboo subject, a dead-end for any career scientist, the next steps in the process die stillborn. This would be a grave undoing of the primal impetus that has brought science from using leeches to cure disease all the way to the unthinkably advanced and almost superhuman feats of our recent history. The three billion base pairs of the human genome have been catalogued, our crafts are at this moment piercing the heliopause on their way into interstellar space, organs are not only being transplanted but the procedures now are considered almost commonplace. Prediction and observation paid for all those great strides forward and will be at the heart of every single step that lies ahead. Without it there would never have been a Galileo or an Einstein; without it science ceases to be science.

A century ago science made a wrong turn regarding seismic forecasting. A small cadre of researchers came to the ill-advised opinion that no one anywhere would ever be able to predict an earthquake. The notion was repeated without challenge for decades and somehow became chiseled in stone. When one asks which titan from humanity's hall of fame first issued the caveat, the silence is deafening. For neither Isaac Newton, or Bertrand

Russell, or Albert Einstein, or any other genius ever said the first word about it. Science must now summon up the courage to admit that—just as with a thousand other supposedly indisputable truths that have gone from undeniable fact to utter falsehoods—a mistake has been made.

Aside from the foundational problems that spring from self-fulfilling prophesies of failure, there is another quite formidable obstacle in getting scientists to agree on a path forward in solving this problem, one having nothing to do with Hellenistic history and very much a modern predicament. I recently met an esteemed mathematics professor in Cleveland while doing a book signing there. He was a very friendly and engaging man and we struck up a conversation concerning a book I'd read about how Andrew Wiles solved the centuries-old problem of Fermat's last theorem in 1995, listed prior to its solution in the *Guinness Book of World Records* as "the most difficult mathematical problem." One of the ways Wiles went at it was using the Taniyama-Shimura Conjecture (which itself has been proved, by the way, and is no "conjecture" now). I'm certainly no great mathematician—hardly! But I couldn't resist the opportunity to wheedle a free tutoring session with the expert at hand. I had made some sense of part of the conjecture but needed a little assistance on the "modular forms" part of the solution and asked the professor if he could help an amateur such as me to understand it better. His answer was unexpected. "Oh, I'm really not that versed in modular forms. From the way you're talking you might know more about it than I do. My specialty is in bifurcation theory." My stunned look told him to continue so he did. "You must realize, my colleagues, mathematics professors, all have about the same grounding in the subject up to first-year graduate-school level, but from there on it's a very highly specialized discipline. All the sciences are that way, but mathematics takes that to extremes."

The dynamics surrounding seismic forecasting involve tremendously varied skills and expertise, requiring the combined collegiality of seismologists, astronomers, mathematicians, physicists, chemists, geologists, and others. In a previous chapter I quoted Dr. Freund's discomfiture regarding how his work had been overlooked because, in his words, geologists weren't interested because of the physics, and physicists weren't interested because of the geology. He's not alone in that opinion.

On August 7, 2013, the *San Francisco Examiner* published my op-ed commentary "California Needs to Take Earthquake Prediction More Seriously" on their opinion pages. It *did* get some interest, not all of it positive. Three much-respected seismologists posted their comments under the piece online, making it very clear that they took umbrage at the piece. I sent the article to Dr. Carlo Doglioni at La Sapienza University in Rome, among the handful of the most esteemed seismologists in Italy, plainly asking where I had gone wrong. In his response to me he echoed Dr. Freund's outlook. "The problem of earthquake prediction being easily solved and celebratory fireworks sent into the air has never been suggested, as you and other fair-minded individuals can see so obviously. You're right about something else too. Finding enough people—world-renowned experts in many fields but yet *with* open minds—to fill an interdisciplinary panel to put this matter right isn't going to be easy. California's Earthquake Prediction Evaluation Council *does* exist, and could have this matter placed before them by the governor, but I, like you, wonder if that will ever be done." That quote comes from the man who discovered that the North American continent was being literally dragged westward by tidal interaction with the Moon, published in the peer-reviewed journal *Nature* in 2006, and featured in *National Geographic* and hundreds of other newspapers and magazines all over the world.

Seismology may want to take Dr. Doglioni's remarks to heart, to wipe the slate clean, brush itself off and start anew—this time with the "open mind" he counsels and in concert with colleagues in a dozen other ancillary fields, which in the end may produce tangible results. Tens of millions in the United States and hundreds of millions around the world are patiently waiting.

San Francisco, April 18, 1906. View northwest from corner of Fourth and Market Streets. (1906-ID. Mendenhall, W.C. 687-mwc00687-US Geological Survey-Public domain image.)

Chapter Sixteen

Vox Populi

In America's democracy it matters little what the scientific community or the government has to say—or refuses to say. Final decisions remain firmly with the people. Italy's Dr. Carlo Doglioni, an expert on how the world is moved by tectonics, in ruminating about the unlikely chances for the people of the United States to make their voice heard regarding their seismic safety, may have underestimated a power every bit as great. In previous chapters at least a dozen promising avenues for tackling the problem of earthquake prediction have been outlined. It's difficult to try and guess which methods might bear real fruit, but it's most probable that a confluence of space-based and ground-based indicators might be what comprises humanity's first seismic alert system. Evidence regarding the existence of credible patterns for how killer quakes have struck in Southern California, the Bay Area, Alaska, and elsewhere has been plainly demonstrated. Unquestionably credentialed scientists—in the United States, Italy, India, Russia, China, Japan—have added their voices to the debate, indicating that a solution is not impossible, and convinced the governments of almost all the seismically active nations on Earth to heed their countenance. And it's not as if the people of the West Coast haven't been apprised of all this. I have spent the last two decades speaking to many, many millions of people—many, many hundreds of times on the air and in print—delivering a clear message that the time to act is now. If the general public is unmoved, if the message has fallen on deaf ears, there's every reason to plainly ask why that is. There might be very good grounds to ignore the clarion call, without casting aspersions on either the messenger or the audience. This is a matter that touches on quite a number of thorny complications.

To start, let's lay out in the plainest terms the nature of the warning. I believe that the next great seismic event in Southern California will most likely occur at either dawn or dusk during either a new or full moon phase date. I believe that the event will transpire in the very near future, not decades from now. It's my view that the authorities may be able to narrow down *which* dawn or dusk by virtue of utilizing the ground-based precursors and indicators outlined in the previous chapters. For the purposes of argument, let's assume that all of the foregoing is correct. Now what? Well, here is what *shouldn't* happen first.

The Moon is either new or full over Los Angeles twenty-four times a year. Obviously, Angelenos can't hunker down in terror owing to unproved "higher probability seismic windows" two dozen times per year. If any population were so balmy as to submit to such abominable interference in their lives, the result would be an unending flirtation with economic and social disaster. Over the years many of these seismic windows have occurred on Christmas, Black Friday, Independence Day—not to mention uncountable millions of personal birthdays, anniversaries, weddings, funerals, graduations, sporting events, etc. Curtailing the shopping habits, celebratory interactions, and day-to-day activities of ten million people across the Southland would not only be impossible, it would be insane, so nothing of the sort is being proposed. That doesn't mean that nothing can be done. What *can* be done might dramatically reduce the casualties and damage should the next great earthquake strike with Los Angeles armed with at least a hint of warning as opposed to being caught absolutely flat-footed. Here, then, are some modest proposals of what *could* be done.

To begin with, it's long past time to set the record straight with regard to higher probability gravitational tidal seismic windows in Los Angeles, once and for all. Either they exist—as I say they do and as I've given proof in the table in chapter five and publicized already on almost every television news channel in Los Angeles (KNBC, KABC, KCBS, KCAL), in ten major newspapers throughout Southern California, on a dozen radio stations across the Southland, as well as the dozen or so times that nationally televised or nationwide radio broadcasts have washed over Greater Los Angeles—or they don't.

The governor's office and the California Earthquake Prediction Evaluation Council have remained mum regarding this matter for the last twenty years, like most government agencies, assuming that the best way to address a potential problem is to wait patiently for it to go away. It hasn't and it won't. The cat has long since been let out of the bag with gutsy Southern California media outlets from Santa Barbara and Ventura to Palm Springs and San Diego who courageously took up the issue and did their job. It's time for the honorable governor and the earthquake council to do theirs.

Popular Science looked at that table in 2013 and said, "This is simple but brilliant observation, followed by reasonable applications; impossible to dismiss as coincidence." KCBS News in Los Angeles spent several weeks in the spring of 2014 going over every iota of the data, calling me repeatedly for clarification when they felt they might have discovered something amiss, demanding that we go online together to access lunar perigee and phase calculators to corroborate line by line every piece of information, insisting that I send third-party confirmation from a mathematics professor that the figures for the overwhelming probabilities were correct, even going so far as to require a letter of verification from the Los Angeles Unified School District attesting to my two decades of service there.

We've all heard the canard that the news media never bothers to do their homework, that they pounce gleefully on anything that will increase their ratings, that they couldn't care less about the truth just so long as they can cook up a scintillating story. I must have very bad luck because I've rarely come across journalists like that, as I've interacted with reporters and editors in every state along with a dozen foreign countries and can count on my right hand the number who fit that description. KCBS finally finished the vetting process and promptly opened their nightly news edition on April 30, 2014 by having anchor Paul Magers asking Angelenos, "What if I were to tell you that nearly every single deadly Southern Californian earthquake in the past happened at dawn or at dusk and during a new or full moon?" Everyone else has had a go at trying to debunk that plain and simple table; perhaps it's time for the honorable governor and the earthquake council to take a crack at it as well. A direct and frank "yes," "no," or "maybe" after weighing the data would be most appreciated, not only by myself, but by every single human being in Greater Los Angeles.

Assuming the verdict is in the affirmative, that first step could reap enormous benefits. Just seeing the dynamic validated by the civil and scientific authorities would provide Angelenos with their first rudimentary protection; they'd have at least a rough idea of when the next great quake might strike. If things were left just there, if nothing else were done, if just the official pronouncement were made, every word in this book would be well worth the ink and paper. Ten million people could then, in the months and years to come, of their own volition, with no heavy-handed safety bureau handing down dictates and directives, go about their business as they saw fit, taking whatever precautions they deemed prudent and feasible for their own families. What might the people of Los Angeles do? Here's the one-word answer: plenty.

Since everyone has a calendar and a clock, millions of people would be able to effect tens of millions of slight precautionary tasks that would require very little effort but in the end, collectively, might make an enormous difference in the event of a large shaker. A couple of dozen times per year Angelenos would have a reminder to look around their homes, to reassess their safety kits and preparations, to perhaps bring down some heavy object temporarily suspended during moving or construction, to move the baby's crib away from a large picture window, to quiz the children about what to do in an earthquake, to fill the car with gasoline, to fill a prescription today and not tomorrow, to make sure slippers are under everyone's beds—the list is endless.

A few years ago a team of Israeli scientists conducted a study of hailstorms on the East Coast and determined, due to the peaks and valleys of pollution emitted on certain days of the week, that Wednesdays were the most probable times for severe hailstorms during winter. New York didn't grind to a halt the next Wednesday, or the one after. Frenzied mobs didn't smash store windows on Tuesday snatching up umbrellas to prepare for the coming disaster. As a matter of fact, instead of the populations of large East Coast cities hiding in fear in the middle of the week or crowding into churches seeking the aid of the Almighty, they simply shrugged their shoulders, put it in the back of their minds, and went about their business as always. Even more surprisingly, not a single word in the press was printed casting aspersions on the study or, more to the point, asking out loud of what use could be made of such an arcane finding. Science doesn't rate new discoveries based on what use they have in the immediate present. It is well known that there is almost no snippet of new information, no novel piece of scientific information that didn't turn out to be the seed of something much greater and more important, even the simple ones.

In a previous chapter a quote was referenced from a very well-known and highly respected functionary of the USGS, given to Reuters, making light of my determination to focus attention on these higher probability seismic windows: "If doing the easy things like the full moon worked we would be doing it." That seems a strong argument at first glance; deeper introspection shines a light on a foundation of flawed logic. Some of the simplest methods have been used to crack open the toughest enigmas in history. Looking for the hardest way around some puzzle, simply because it has proved difficult to tackle in the past, gives no assurance that such a route is the correct one. Sometimes the correct path *is* the easy path. Our understanding of lightning serves as an excellent example.

No other natural phenomenon would have frightened our ancestors like lightning. It has a particular terror all its own. Hurricanes, tornadoes, wildfires, avalanches, volcanic explosions, and all the rest either can be seen coming or develop their killing potential over a few moments, a few hours, a few days. Nothing strikes so suddenly, so quickly, so terrifyingly as lightning. And while an earthquake shakes an entire region, lightning pinpoints its focus of power on a single spot. No wonder the ancients imagined it being thrown purposefully by the gods. In our modern era, while appeasing Zeus has been crossed off the list of the best ways to avoid being struck, there are a few expedients still at hand. Here's the surest—and the simplest—one of them all: go down. Being warned to come down from higher elevations during a lightning storm is sage advice, and a recommendation that is given by any and all safety councils the world over. Who was the genius that discovered such an important, practical, beneficial, yet simple safeguard? His or her name is not among the many eminent scientists referenced in this book, or any other. He or she is unknown to history. Notwithstanding both its lowly pedigree and unpretentiously straightforward method, this truth about lightning is nonetheless heeded, and it is only the foolish and the daredevils who climb to the tops of mountains rather than hunker down in valleys during thunderstorms.

Benjamin Franklin should be counted among the elite group of the greatest thinkers to have ever lived. In addition to bestowing his political wisdom on the American people, this Founding Father also gave mankind its first real defense against lightning. It's hard to imagine the extent to which lightning held colonial America in its thrall. So many fires were caused by this phenomenon that church bells would ring all across the colonies whenever storm clouds approached. To realize that such an ancient peril could be brought to heel by the easy expedient of affixing a lightning rod to structures—nothing more than an iron shaft—is proof positive that unadorned, down-to-earth measures are not to be dismissed out of hand. Franklin made sure that no patent was ever issued for his invention, deeming the device too important to the world for any personal monetary gain.

Storms at sea even today are enough to cause the knees of the bravest sailors to weaken. During the Age of Sail, with men challenging the oceans in rickety wooden vessels that were hardly proof against such power, the ferocious energy of the sea was sufficient to bring many a sailor to the brink of endurance and beyond. In the days before radar, weather satellites, meteorological forecasts, and ship-to-shore radio, the only real insurance that a ship would arrive at its destination was the bravery and competence of its crew and the luck of the draw. Sharp-eyed seamen came to rely on a certain practical early warning measure where none other existed. "Red sky in morning, sailor take warning; red sky at night, sailor's delight" is not only poetic—it's true, at least in the Northern Hemisphere. Shakespeare makes mention of the dictum in his poem *Venus and Adonis*: "Like a red morn that ever yet betokened, wreck to the seaman, tempest to the field, sorrow to the shepherds, woe unto the birds, gusts and foul flaws to herdsmen and to herds." Jesus Christ Himself is quoted in the Bible (Matthew 16: 2–3) as advising "When in evening, ye say, it will be weather, for the sky is red. And in the morning, it will be foul weather today, for the sky is red and lowering." The Sun's rays, at dusk and dawn, shine through the thickest part of the atmosphere and are subject to changes in their color spectrum due to droplets of water in storm clouds that might interpose themselves between the observer and the Sun. Since weather in the Northern Hemisphere generally moves from west to east, this phenomenon can give some indication if a storm is on its way over the horizon or if it is already past the observer's position.

It's a great fiction that people in Columbus's time thought that the world was flat. Certainly the uneducated thought that was the case, but knowledge of the true shape of our planet goes back to the mists of antiquity. Pythagoras, Aristotle, Ptolemy, and uncounted other scholars of the distant past wrote as much. Probably the first to document this fact were Babylonian astrologers from around 1000 BC, although the knowledge probably predates them too. The complicated mechanism they used to elicit such a discovery was nothing more than their own eyes. During eclipses Earth's rounded shadow is cast upon the Moon. Also, the tall masts of ships appear first over the horizon before an approaching vessel itself is seen, confirming Earth's curvature. Eratosthenes proved it in Alexandria by using nothing more than the shadow of a stick and a deep well.

Einstein would certainly have remained in his post as a Swiss patent clerk, but for the work of his intellectual predecessor, Sir Isaac Newton. And Newton himself—hardly the paragon of humility—nevertheless admitted that he only accomplished what he did by "standing on the shoulders of giants." It is obvious to whom he was referring: Johannes Kepler, Tycho Brahe, and before all of them—Galileo Galilei. Our view of the universe rests, in Newton's words, upon the shoulders of this incomparable giant. As pointed out in a previous chapter, Galileo made his monumental discoveries without the use of a supercomputer, unaided by a cadre of research assistants, using the relatively simple mathematics prior to calculus, absent corroborating emails from scientific academies around the world. His tools were primitive enough to have been utilized by a Cro-Magnon "scientist": two stones dropped from a height—one heavier than the other. Those two weights gave birth to everything we know about the universe because their thud upon the ground reverberated through the centuries into the mind of Einstein.

Researchers all over the world are currently engaged in the final great task to bring everything in science full circle, searching for a Grand Unified Theory—a so called "theory of everything," which would unite quantum mechanics and relativity. They're using something more than sticks and stones; CERN's stupendous Large Hadron Collider (a tunnel twenty-seven kilometers in circumference under the Franco-Swiss border), which can smash protons together at just a fraction below the speed of light, might just do the trick. They're not ruling anything else out, though. The greatest minds alive today are all of the opinion that when they do see the theory they're looking for it will have a hallmark, an attribute that they are sure will qualify it as being true and correct. It will be simple and elegant. When the universe speaks truly, it always speaks in like fashion.

There's no reason to demand that anything to do with earthquake prediction must be accompanied by reams of tensor calculus equations. That would be to discount the plain empirical evidence that Earth has always generously provided and that in previous ages allowed us to gain a toehold on countless other enigmas. That the ideas within this book are unadorned, austere, and unpretentious—that they are "easy"—might be the best indication for their being . . . true.

And if that be the verdict of the California Earthquake Prediction Evaluation Council, aside from what the general populace can do on their own, there are many, many other very simple, practically cost-free expedients at hand should the civil and scientific authorities lend their weight to potentially greatly ameliorating the next great earthquake. Regarding the expense of eliciting a judgment from that august body, it couldn't cost an exorbitant amount to convene the Earthquake Council for the purposes of examining the data set out in the table of great earthquakes strong enough to have caused fatalities in Greater Los Angeles in chapter 5 and to deliver a conclusion. Any of the members too busy with other matters might select a proxy from among thousands of qualified experts in California. Following Dr. Freund's gracious and civic-minded example of putting up his own funds (over $1 million) in order to nudge NASA in what he considers the right direction, I would be more than pleased to finance the endeavor myself. Paying heed to the ground-based precursors outlined in previous chapters, in conjunction with the times and dates indicated by gravitational tidal triggers, could result in something the West Coast lacks yet sorely needs, something China has had for half a century and Japan for more than a decade: a rudimentary earthquake warning system. It wouldn't be foolproof, but it would be a start. Here is how it might work.

Let us assume that some ominous precursors have been noted around Southern California. It could be that helium-3 is starting to vent with a vengeance along the Newport-Inglewood fault, or a radical change in electrical resistivity is logged along a section of the Mojave or Coachella segment of the San Andreas. Let us also assume that in two or three days there will be a somewhat rare full-moon and perigee event, pushing tidal stresses to their absolute maximum. In China the government has evacuated areas in similar situations; Los Angeles can't be emptied, however. So, rather than pushing any panic button, there are simple, sane, safe measures that might be taken, without even informing the general public that anything was being done at all.

Police, fire, EMS, National Guard, and other agencies could be given a simple heads-up—nothing more than a communiqué. Perhaps the rosters of those first responders could be changed ever so slightly—a few additional police and firefighters on call, mention at least being made by supervisors as shifts go out to duty. The San Onofre nuclear power plant could make preparations to avoid the terrible fate of reactor meltdowns at Fukushima following the Japanese quake and tsunami.

It all sounds exceedingly simple, doesn't it? One can almost be lulled into wondering how it is that no one has suggested anything like it before. Well, Dr. Carlo Doglioni's email to me casts the proper light on the foregoing. He declared that he realized that I was on record as stating that this is anything but an undemanding problem, that I never intimated that "fireworks should be sent up" to accompany the hosannas offered up to mark the solution of one of the world's great conundrums. Carlo is absolutely right; this problem is far from solved.

Let's rewind our hypothetical scenario, and this time let's see what happens in the real world. Alarm bells are ringing due to massive outgassing or some other precursory indicator. As per the plan laid out above, and in order that panic not ensue, only the first responders are sent communiqués advising that an alert is in effect. The LAPD sergeant dismissing the next corps of officers to take to the streets advises them of the fact, accompanied by some grumbling by those who are going to pull overtime because of it. It is a surety that the first thing those officers will do as they exit the station house will be to reach for their cell phones. Every officer will call their spouse, children, relatives, friends, stock broker, minister or priest or rabbi. Every firefighter, EMS paramedic, and engineer at Pacific Gas & Electric will do the same. Within minutes Los Angeles will be abuzz not only with news of the possibly impending event, but worse, it will be cloaked in secrecy, the very thing to put minds so far toward unease as to create exactly what it was intended to prevent: panic. A thorny no-win scenario now takes place. If no earthquake happens, if it turns out to be a false alarm, those injured in the hectic shuffle to prepare for disaster or who claim to have been subjected to mental anguish for no reason might well appear the next day at city hall accompanied by their attorneys. If an earthquake *does* strike, it will be common knowledge that the city was left uninformed save for a very few, and anyone injured or the relatives of those killed will most definitely be screaming for justice.

So let's give another scenario a test run. Perhaps it won't do to only inform first responders, city officials, and a select few. Maybe the only way forward is to bite the bullet and issue a bulletin to be broadcast on all news and social media. The lawyers would love this option. Los Angeles might suffer greater financial harm from the torrent of lawsuits than from an earthquake if no tremor is felt—for among the hordes of supplicants demanding redress would be some heavy hitters indeed.

In chapter 9 I told the distressing tale of Dr. Whitcomb, the Cal Tech researcher who was pilloried by real estate brokers and city officials for his faulty appraisal of an impending earthquake. The problems he encountered would shrink into insignificance compared to what the great retailers and other business leaders of Southern California would demand, for as Angelenos dropped everything and made for their homes and families during the alert their losses would be real and perhaps quite staggering. At end, as the legal ramifications played out, the punishments and fines levied might make the L'Aquila earthquake affair pale in comparison.

So what *is* the answer? Well, Dr. Doglioni gave it already: there is none at present. What is going to be required can't be found in this book or any other. Rather than engaging in paroxysms of handwringing about why Californians haven't jumped willy-nilly toward the idea of a seismic warning system and demanded its implementation, it might be well to remember that the collective will of the people is something that only very foolish individuals assume to be faulty, and their personal opinion the correct view. It may be that deep down there is a realization that allowing seismic assaults to remain quietly undisturbed in the "act of God" category, for now at least, might be the best option from among a very bad slate of choices.

To change this, some towering figure like William Mulholland from Los Angeles's past is going to have to step forward and by the force of his or her personality wrestle with it until the problem is ground down by sheer stubbornness and willpower. Mulholland faced the same kind of insoluble dilemma from 1902 to 1913 and the saga of how he brought water to a city desperate for the resource is an epic to rival any other. The chronicle includes a shooting war with Owens Valley farmers who fought unsuccessfully to prevent their water from being exported to Southern California, a failed dam that drowned almost five hundred people in the Santa Clara River Valley, and the construction of an aqueduct that stands with any engineering monument on Earth. When the California Aqueduct was completed, and Mulholland stepped up to give the dedication speech to the crowd of Angelenos gathered at the terminus of the channel, he casually pressed the button detonating the last barrier holding back the water. As the cascade roared down the causeway he turned to the audience and gave the shortest inaugural speech in history: "There it is, take it." It's unknown if they make men like him anymore, but Los Angeles will need someone very much like him nonetheless.

Whatever body is chosen to stand guard over Los Angeles's seismic safety, when and if that should happen, its leadership is going to require a sage and courageous director, whether it be the California Earthquake Prediction Evaluation Council, the United States Geological Survey, or some other. Actually, Los Angeles could do far worse than the Southern California Earthquake Center. Its director, Dr. Tom Jordan, is eminently competent, honest to a fault, absolutely fearless of public opinion, completely open minded, and one of the few high-ranking functionaries whose words don't change whenever a microphone is placed before him. His is the face America saw on national television for untold hours on end during the Japanese crisis in 2011. When the L'Aquila seismologists were arrested, he was the only nationally recognized seismologist to put the entire affair in the proper perspective while the rest of the scientific world was too busy engaged in a cacophony of invectives to see both sides. While he, of course, disagreed strongly with the absurd decision, he was a big enough man to give voice to what motivated the prosecution. "People are expecting much more information, in particular quantitative information. Coming clean with what you know is being demanded by the public," he told the

New York Times in 2011. I had never heard a top-tier seismologist utter such words, at least not in the United States. But then he's the same man who told AOL News in 2010, concerning a previous book of mine, that yes, science is aware that "small quakes" at least are known to be triggered by gravitational tides, just that the jury was out regarding the bigger ones. That matter-of-fact candor contrasts strongly with the previous fourteen years of tongue-tied convolutions pronounced by others. So he may not measure up to William Mulholland, but he might well do in a pinch in any event.

The Russian astronomer Nicholai Kardashev categorized the three types of technological civilizations that could exist in the universe. A Type I society would be a race that had risen to the stage of harnessing the total energy of its home planet. Type II cultures progress to the level of using all the energy of their host star. Type III civilizations would have advanced to colonize an entire galaxy, mastering the combined energy of a hundred billion stars. A Type III civilization would be capable of absolutely anything: deconstructing and reconstructing planets, turning stars on and off, moving black holes around like playthings, creating exotic matter and using it to build wormholes, perhaps even birthing baby universes. It would be the most powerful single thing in the universe. There are no Type III or Type II cultures anywhere in this vicinity, but there is a very good candidate civilization currently knocking at the door of Kardashev's Type I level. It's the society living on the third planet from a G-type main sequence star about two-thirds of the way out in one of the Milky Way galaxy's spiral arms.

As the centuries pass and mankind's control is established over every facet of Earth, forthcoming generations could plausibly regard the future residents of California as particularly fortunate to live next to the San Andreas Fault—that titanic generator of such abundant electricity. We're not that far away from such descriptions of the human race. It's an almost equidistant journey in time forward to the brink of the third millennium, or back to the signing of the Magna Carta during the High Middle Ages. That wasn't really so long ago. Going back even further, to the beginning, to Sumer and Egypt in 3000 BC, the picture is still far from alien. Sumerians read and wrote, did math, paid taxes, shopped in markets, obeyed laws. Egyptians worried about fashion, put on makeup, surveyed their farms, worshiped on holy days, threw parties, and got drunk. That wasn't really so long ago. Five thousand years from now—in the year 7000—the germ of life that sprouted on this planet billions of years ago will have matured into the single most powerful force in the cosmos: an immortal, godlike race that will hold the entire Milky Way in its hands. Such a Type III civilization would be able to slip a leash around the San Andreas with hardly more effort than what's required for us to imagine such things now.

In the distant future, no matter how advanced our descendants become, no matter what superhuman feats they'll be able to accomplish with ease, they will doubtless look back on their ancestors in early twenty-first century with deep pride and boundless admiration. For with the primitive tools at our disposal, with skills barely equal to the task, we will nonetheless have courageously taken on a supremely formidable opponent, and like our own forebears in countless other terrific battles against nature, conquered. It can and will be done—because it must.

Chapter Seventeen

Earthquake Preparedness

Before the Next Great Earthquake

James Hubbard, "the Survival Doctor," is quoted as saying that "everyone deserves the chance to survive and every time I see another disaster I think that there are people who are dying who don't have to." The civil authorities in seismically active regions all over the United States over the last decades have made Herculean efforts in retrofitting structures, modifying building codes, holding regular citywide and statewide disaster preparedness dry-runs, establishing action plans and command centers, and a host of genuinely admirable and eminently useful endeavors to prepare for the next great earthquake. There is nothing, however, that can take the place of each and every citizen doing his and her small part—through simple and basic earthquake preparedness—to ensure that no matter what the magnitude of the next crisis, it is mitigated as best it can so that people don't die who don't have to.

- Secure bookcases, china cabinets, entertainment units, large picture frames, mirrors, and the water heater. Using angle brackets and strapping, make sure to drill or screw into the studs in the walls. Where possible, attach wooden strips of molding across open shelves to prevent contents from falling.
- Where are the shutoffs for your gas, water, and electrical utilities? Find them and make sure everyone in the family knows how to turn them off. Make sure the tools to accomplish this are nearby.
- Is your home supported stably on its foundation? If you're not sure and you're not a contractor or an engineer, have it inspected. Also, check brick facing, chimneys, and retaining walls that may require buttressing to withstand a severe jolt.
- Make a family plan. Make sure the children and elderly in your household know what to do in the event of a severe earthquake. Determine in advance how your family will reunite if the quake strikes when you're separated. Make sure everyone has a list of contact phone numbers. Designate an out-of-state contact that you can use to relay messages. Store emergency supplies. Practice ducking, covering, and exiting with your small children.
- Make a community plan. Who in your neighborhood has the garage full of tools? Are there doctors, nurses, firefighters, or other people living on your block with special skills and equipment? By pooling talents and resources you can make preparations that will mitigate even the worst-case scenarios.

- Learn CPR.
- Make sure to keep on hand a plentiful supply of any medications such as insulin or heart pills that are crucial in preventing a life-threatening situation.
- Try to keep a full tank of gasoline in the car.
- Keep flammable and hazardous materials someplace safe, somewhere close to the ground.
- Don't hang heavy objects above sofas or beds.
- Mirrors, pictures, and other objects on walls should be hung with closed hooks.
- Check and replace rusted or worn water and gas pipes.
- Alert children to areas in every room where they can take cover (desk, tables, etc.).
- Keep flammable and hazardous materials someplace safe, somewhere close to the ground.
- Shoes and flashlights should be kept under every bed.
- If your home has rigid gas connections to water heater, stove, or dryer, have a plumber replace them with flexible, stainless steel corrugated connectors.
- Excess gas flow cutoff valves are available to stop a catastrophic leak.
- Store emergency supplies (see below).

Emergency Supplies

There may be few resources available to manage the day-to-day contingencies of life after a major seismic event, so it is important to plan ahead and have supplies on hand.

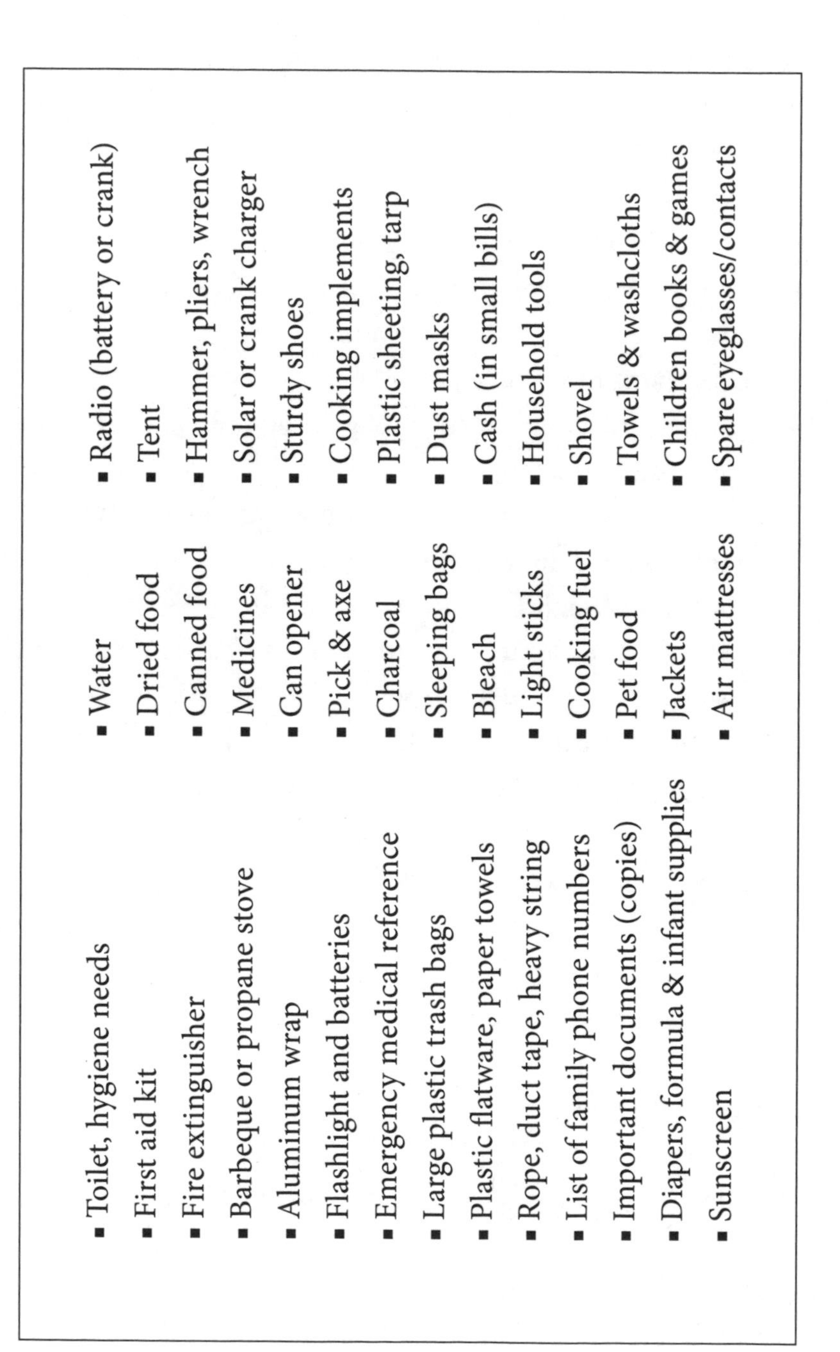

- Toilet, hygiene needs
- First aid kit
- Fire extinguisher
- Barbeque or propane stove
- Aluminum wrap
- Flashlight and batteries
- Emergency medical reference
- Large plastic trash bags
- Plastic flatware, paper towels
- Rope, duct tape, heavy string
- List of family phone numbers
- Important documents (copies)
- Diapers, formula & infant supplies
- Sunscreen

- Water
- Dried food
- Canned food
- Medicines
- Can opener
- Pick & axe
- Charcoal
- Sleeping bags
- Bleach
- Light sticks
- Cooking fuel
- Pet food
- Jackets
- Air mattresses

- Radio (battery or crank)
- Tent
- Hammer, pliers, wrench
- Solar or crank charger
- Sturdy shoes
- Cooking implements
- Plastic sheeting, tarp
- Dust masks
- Cash (in small bills)
- Household tools
- Shovel
- Towels & washcloths
- Children books & games
- Spare eyeglasses/contacts

Storing and Purifying Water

Access to clean water must be a priority in emergencies. An earthquake can rupture water mains and disrupt electrical power needed for pumping and filtration. Make sure you have enough safe water on hand for drinking, cooking, and cleaning.

- Store water in a cool, dark place in plastic containers. You can even use two-liter soft drink bottles if you like. Milk and juice containers don't work as well as they have more of a tendency to crack and leak. Keep track of the date and replace the water every year at least.
- Use liquid household bleach (sodium hypochlorite 5.25%) to insure a safe, potable water supply. Add two drops of bleach per quart stored, or eight drops per gallon, one half teaspoon per five gallons, and five teaspoons per fifty-five gallon drum. If your faucet water is particularly cloudy, increase the bleach ratio by an additional 50%.
- Don't forget to store some water in your car.
- In an emergency, there are many sources of potable water in your home or apartment. You can drink the water in the toilet reservoir tank, and also utilize the supply drained from the water heater faucet, if the water heater hasn't been damaged.
- The swimming pool may have to serve as a bathtub, but you should not drink that water.
- Don't drink the water from automobile radiators or waterbeds.
- If water has to be boiled, do it for a full five minutes; at higher elevations keep it at a full boil for ten minutes.

- Melted ice cubes, canned fruit, vegetable juice, and liquids from other canned goods are other sources of water.

During the Next Great Earthquake

Panic has always been the great killer in every historic earthquake. The single most important defense against injury or death in an earthquake cannot be found in any preparedness kit. It isn't enough to know and follow the strictures below; first and foremost is to keep one's head.

- Drop, cover, and hold on under a table, desk, or other sturdy protection. If you are unable to accomplish that at least distance yourself from windows or heavy objects that might fall. Standing in the nearest doorway might have been a good idea in 1857, when many homes in California were constructed of adobe and the sturdy yet flexible door jambs might have held even as the bricks were shaken down. It is inadvisable now. Brace yourself. Protect your head with your arms and wait for the shaking to stop.
- If you are in a high-rise building, do not use the elevator to exit after the shaking stops. Go out through the stairway.
- If you're outside during an earthquake, that's probably a good thing. Just use common sense and move away from anything that might fall—especially power lines.
- If you're driving when the next quake hits, pull your car over and stop as soon as that can safely be done. Stay inside your car.

- Prepare for aftershocks. If it's a big earthquake there will be aftershocks. Get yourself to the nearest, safest place possible.
- If you live near to the shoreline there is always the threat of tsunamis produced by great seismic disturbances. It's safest to avoid approaching the seaside until several hours after the shaking stops. Obviously, if one is at the beach when an earthquake strikes it is advisable to seek higher ground.
- Those living below dams should rightly be concerned about catastrophic failure if the earthquake is a particularly strong one. It is prudent to consider leaving the immediate area until it has been confirmed by authorities that the dam in question hasn't been compromised.

After the Next Great Earthquake

Every great city that recovered from a calamity did so via an integral and organic collective will to put civic recuperation at a higher level. When the shaking finally stops is when each citizen should begin to think of himself not as an individual survivor, but as part of a united civilian team capable of putting right even the worst situations. There is no fault line, no power on Earth, either on its surface or below it, that can undo a united citizenry working together.

- Check for fires, gas leaks, electrical damage, and any other urgent contingency requiring immediate attention. Stay away from brick chimneys and other high-placed masonry construction. It's also best to keep one's distance from power lines.

- Obviously, apply first aid to anyone hurt. Those with very serious injuries shouldn't be moved and should be treated and comforted in situ while emergency services are awaited.
- Inspect your home. If it has suffered serious damage and is unsafe, evacuate it. Post a notice informing interested parties where you can be found.
- Don't use the telephone needlessly. The lines need to remain free for emergency communications and cell and land line systems could be jammed. Of course, report any life-threatening emergencies. Text messages take less bandwidth and may go through when voice calls are failing.
- Don't use your vehicle needlessly. In the event of a severe earthquake, the city's infrastructure will be damaged—roads, power lines, sewers, water, gas. Stay put.
- Check with your neighbors. Lend assistance where you can.
- Water and/or sewer lines may be broken. If they are you may need to consider digging a latrine and lining it with plastic bags for removal later.
- Turn on your radio (or TV, if possible). Listen for official instructions. There may be a curfew or other mandates in effect.
- Prepare for aftershocks. If the earthquake is a strong one there almost certainly will be aftershocks, potentially very strong ones. Moreover, the initial earthquake may have been a foreshock, meaning that the strongest jolt is yet to come.

There may be some very inconvenient problems in the days and weeks after a large earthquake, and it is prudent to give some thought to the following list of possible situations.

- Water service may be out; gas and electric also could conceivably be suspended for days or weeks.
- Garbage and sewer services may be nonexistent.
- Telephone, Internet, cell phone, cable, wireless communication may be overloaded, intermittent, or lost completely. Even US mail delivery could be affected.
- There might very well be gasoline rationing.
- With ATMs and online banking possibly disrupted, access to cash may be limited.
- Groceries, pharmacies, and other retail stores may be unable to restock their shelves due to transportation difficulties and other obstacles. Food, medicine, and other necessities may be more difficult to find depending on the specific items.
- Payrolls may be impossible to meet or disburse; family income may cease for a time.
- Earthquakes can be hard on children. After a severe earthquake it's possible for a child to develop problems that might be the result of such a disruption in their environment. The physical symptoms include diarrhea, constipation, headaches, nausea, sleep disorders, loss of appetite, overeating, and skin rashes. Changes in behavior are also a possibility: hyperactivity, nervousness, disobedience, irritability, inability to concentrate, depression, and withdrawal. Time should cure any of these ills, but if problems persist one should seek professional help.

Resources

American Red Cross
http://www.redcross.org/services/disaster/
1-866-GET-INFO (438-4636)

Govenor's Office of Emergency Services (OES)
http://www.oes.ca.gov/

Federal Emergency Management Agency (FEMA)
http://www.fema.gov/about/process/

United States Geological Survey
http://earthquake.usgs.gov/learn/preparedness.php
USGS Southern California Earthquake Preparedness Information: http://www.earthquakecountry.info/roots/
USGS Northern California Earthquake Preparedness Information: http://pubs.usgs.gov/gip/2005/15/
USGS General Information Product 41, "Protecting Your Family From Earthquakes—The Seven Steps to Earthquake Safety" (English and Spanish)
USGS General Information Product 42, "Protecting Your Family From Earthquakes—The Seven Steps to Earthquake Safety" (in English, Chinese, Vietnamese, and Korean)

Earthquake Country Alliance (Southern California)
http://www.earthquakecountry.org/

Bay Area Earthquake Alliance (Northern California)
http://bayquakealliance.org/prepare/

Oregon Office of Emergency Management
http://www.oregon.gov/OMD/OEM/Pages/plans_train/earthquake.aspx

King County Emergency Management (Seattle/Tacoma)
http://www.kingcounty.gov/safety/prepare/residents-business/Hazards_Disasters/Earthquakes.aspx

Emergency Management (British Columbia)
http://www2.gov.bc.ca/gov/content/safety/emergency-preparedness-response-recovery

State Emergency Management Agency (Missouri)
http://sema.dps.mo.gov/earthquake_preparedness/

Tennessee Emergency Management Agency
http://www.tnema.org/public/otherthreats.html

Appendix

List of Full and New Moon Phase and Perigee Dates (2017-2025) (Times are listed in universal time, GMT. Pacific Time is 8 hours earlier; 7 hours earlier during DST. Examples: Jan 10, 2016 1:31 converts to Jan 9, 5:31 PM; Feb 8, 2016 14:40 converts to Feb 8, 6: 40 AM)

New	Full
2016 Dec 29 6:54	2017 Jan 12 11:35
2017 Jan 28 0:08	2017 Feb 11 0:34
2017 Feb 26 15:00	2017 Mar 12 14:55
2017 Mar 28 2:59	2017 Apr 11 6:09
2017 Apr 26 12:18	2017 May 10 21:44
2017 May 25 19:46	2017 Jun 9 13:11
2017 Jun 24 2:33	2017 Jul 9 4:09
2017 Jul 23 9:47	2017 Aug 7 18:13
2017 Aug 21 18:32	2017 Sep 6 7:05
2017 Sep 20 5:31	2017 Oct 5 18:42
2017 Oct 19 19:13	2017 Nov 4 5:24
2017 Nov 18 11:43	2017 Dec 3 15:49
2017 Dec 18 6:31	2018 Jan 2 2:25

Perigee 2017		
1/10 6:08	2/ 6 14:00	3/ 3 7:25
3/30 12:40	4/27 16:19	5/26 1:24
6/23 10:50	7/21 17:11	8/18 13:16
9/13 16:05	10/9 5:52	11/6 0:10
12/4 8:43		

New	Full
2017 Dec 18 6:31	2018 Jan 2 2:25
2018 Jan 17 2:18	2018 Jan 31 13:28
2018 Feb 15 21:07	2018 Mar 2 0:52
2018 Mar 17 13:14	2018 Mar 31 12:38
2018 Apr 16 2:00	2018 Apr 30 0:59
2018 May 15 11:50	2018 May 29 14:21
2018 Jun 13 19:45	2018 Jun 28 4:55
2018 Jul 13 2:50	2018 Jul 27 20:22
2018 Aug 11 9:59	2018 Aug 26 11:58
2018 Sep 9 18:03	2018 Sep 25 2:55
2018 Oct 9 3:48	2018 Oct 24 16:48
2018 Nov 7 16:03	2018 Nov 23 5:41
2018 Dec 7 7:22	2018 Dec 22 17:50

Perigee 2018		
1/1 21:56	1/30 1:55	2/27 14:50
3/26 17:19	4/20 14:26	5/17 21:07
6/14 23:56	7/13 8:30	8/10 18:06
9/8 1:23	10/5 22:31	10/31 20:06
11/26 12:11	12/24 9:53	

New	Full
2019 Jan 6 1:30	2019 Jan 21 5:17
2019 Feb 4 21:05	2019 Feb 19 15:54
2019 Mar 6 16:05	2019 Mar 21 1:43
2019 Apr 5 8:52	2019 Apr 19 11:12
2019 May 4 22:47	2019 May 18 21:12
2019 Jun 3 10:03	2019 Jun 17 8:31
2019 Jul 2 19:17	2019 Jul 16 21:40
2019 Aug 1 3:13	2019 Aug 15 12:31
2019 Aug 30 10:38	2019 Sep 14 4:35
2019 Sep 28 18:28	2019 Oct 13 21:11
2019 Oct 28 3:40	2019 Nov 12 13:37
2019 Nov 26 15:08	2019 Dec 12 5:15
2019 Dec 26 5:16	2020 Jan 10 19:23

Perigee 2019		
1/23 19:59	2/19 9:07	3/19 19:48
4/16 22:03	5/13 21:54	6/7 23:23
7/5 4:56	8/2 7:10	8/30 15:59
9/28 2:28	10/26 10:42	11/23 7:56
12/18 20:31		

New	Full
2019 Dec 26 5:16	2020 Jan 10 19:23
2020 Jan 24 21:44	2020 Feb 9 7:35
2020 Feb 23 15:34	2020 Mar 9 17:49
2020 Mar 24 9:30	2020 Apr 8 2:36
2020 Apr 23 2:27	2020 May 7 10:46
2020 May 22 17:40	2020 Jun 5 19:13
2020 Jun 21 6:42	2020 Jul 5 4:45
2020 Jul 20 17:34	2020 Aug 3 16:00
2020 Aug 19 2:42	2020 Sep 2 5:23
2020 Sep 17 11:01	2020 Oct 1 21:07
2020 Oct 16 19:32	2020 Oct 31 14:51
2020 Nov 15 5:09	2020 Nov 30 9:32
2020 Dec 14 16:19	2020 Dec 30 3:30

Perigee 2020		
1/13 20:22	2/10 20:32	3/10 6:34
4/7 18:10	5/6 3:05	6/3 3:38
6/30 2:10	7/25 4:55	8/21 11:00
9/18 13:45	10/16 23:48	11/14 11:49
12/12 20:43		

Acknowledgments

A great debt of thanks is owed to hundreds of editors, producers, columnists, radio and television show hosts and anchors, scientists, reporters, government officials and many others without whose gracious assistance this book would have been impossible. It's not feasible to credit every individual who helped bring the work to fruition, but to all the individuals, organizations, and news outlets from all over the world listed below, and those whose names are unfortunately missing, you have the sincerest thanks of me and also of the people of West Coast whom you have served splendidly. Many of the people, news venues, and agencies listed below have supported the idea of a more thorough investigation of seismic matters. Others have only judged the issue worthy of placement upon the public forum and as a result done their civic duty, or been in the leadership position of media sources that have opted to bring this matter before their readers, listeners, and viewers. So nothing is implied with regard to what must be hundreds of different shades of opinion. I am most grateful to the many, many thousands of people who have taken the time and trouble to post comments and send letters to editors weighing in on this subject and to the hundred or so blogs far too numerous to list who have made mention as well—either pro or con. In the end it is those voices that matter most.

Danielle Elliot, CBS News National Science Desk; Thom Hartmann, *RT Television*; Alan Taylor and Dr. Cort Stoskopf, *Popular Science*; Howard Stern and Robin Quivers, *Howard Stern Show*; Matt Drudge, *Drudge Report*; Deborah Norville, *Inside Edition*; George Noory, *Coast to Coast AM*; Andrew Ireland, *World Net Daily*; Paul Dacre, editor, *London Daily Mail*; Joe Madison, MSNBC contributor/Sirius Radio Network , *Joe Madison Show*; Mariel Garza, opinion page editor, *Los Angeles Daily News*; Steven Buel, editor, *San Francisco Examiner*; Debbie Henley, editor, *Newsday*, New York; Erin Aubrey, *Los Angeles Weekly*; *Politics Daily*; Dr. Friedemann Freund, NASA, SETI Institute, San Jose State University; Dr. Carlo Doglioni, La Sapienza University, Rome; Dr. Vinayak Kolvankar, chief seismologist (retired) Seismology Division, Bhabha Atomic Research Center, Mumbai, India; Dr. Marvin Herndon, University of California, San Diego; David Ono, anchor, *KABC-TV News*, Los Angeles; Professor Kristian Schlegel, History of Geo-and-Space Sciences, Gottingen, Germany; Conan Nolan, anchor, *KNBC-TV News*, Los Angeles; Paul Magers, anchor, *KCBS-TV News*; David Page, *KSRO Morning News*, San Francisco Bay Area; David Cruz, *David Cruz Show*, KFI, Los Angeles; Kevin Ryder and Gene "Bean" Baxter, *Kevin and Bean Show*, KROQ, Los Angeles; Shane Harrison, host, *Irish Side of the Moon*, Derry, Northern Ireland; Dr. Kate Hutton, chief seismologist, California Institute of Technology, Pasadena; Hillary Frey, editor-in-chief, Yahoo News; Larry Carroll, anchor, KCAL-TV, Los Angeles; Scott Cox, KERN-TV and radio, Bakersfield; Emmitt Miller and Dana Adams, cohosts, *Strange Universe*, UPN-TV Network; John "Stuttering John" Melendez, producer, *Howard Stern Show*; Rudy Ugas, *Despierten Magazine*, Caracas, Venezuela; Patricia Adams, executive director, *Probe International Magazine*, London; Dr. Chris Smith, *Naked Scientists*,

London; Lou Penrose, *Lou Penrose Show*, KNWZ, Palm Springs; Rob Mc Connell, *The 'X' Zone*, Toronto; Whitley Strieber, *Dreamland Radio Show* and *Above Top Secret*, Los Angeles; Mike Craft, opinion page editor, *Ventura Star*; Bill Leff and Wendy Snyder, *Bill and Wendy Show*, WGN, Chicago; Peter Price, *Beyond the Dark*, Radio City, Liverpool, UK; Justin Sacher, *CBS-TV Eyewitness News*, Fresno; David Highfield, KDKA-TV, Pittsburgh; Professor David Swigon, University of Pittsburgh; Larry Richert and John Shumway, coanchors, *KDKA Morning News*, Pittsburgh; Gina Keating, Reuters; Don Katich, director of news operations, *Santa Barbara News-Press*; David Moye, *America Online News*; Kate Delaney, *America Tonight*; Doug McIntyre, *McIntyre in the Morning*, KABC, Los Angeles; *Mind over Matters, Arrow 96*, Los Angeles; Tony Landucci, news producer, CNN/KSRO; Branden Rathert, *Pensacola Now*, NewsRadio 1620, Pensacola; Mike Lonergan, *Lonergan Program*, KLAY, Seattle/Tacoma; Alice Rowley, opinion editor, *Pittsburgh Post-Gazette*; Frank Craig, editor, *Pittsburgh Tribune-Review*; Bridgetta Tomarchio and Sam Hasson, *Sam in the Morning, LA Talk radio*; Michelle Meow, *Swirl Radio*, Oakland and Seattle; Jonathan Dean, producer, *City Talk*, Liverpool, UK; Michelle Jackson, *Michelle Jackson Show*, KRXA, Monterey; Dr. Molly McClain, professor and editor, *Journal of San Diego History*; Professor Paolo Palmieri, University of Pittsburgh; Rand Morrison, senior producer, *CBS News' Public Eye with Bryant Gumbel*; Professor Theodore Arabatzis, History of Science Department, University of Athens; Frank Gerardo, editor, *Pasadena Star-News*; Bill Bell, editor, *Whittier Daily News*; Dorothy Reinhold, editor, *San Gabriel Valley Tribune*; Adrian Mata, constituent affairs director, Office of the Governor of the State of California; Michael Anastasi, editor, *Long Beach Press-Telegram*; Marc Amazon, *Marc Amazon Show*, WLW,

Cincinnati; *Torrance Daily Breeze; Merced Sun-Star*; *Bakersfield Californian*; Joe Buda, publisher, *Las Vegas Informer*; Don Swift, publisher, *ion Oklahoma Magazine*; Jeff Jardine, columnist, *Modesto Bee*; Dave Bowman, show host, *Afternoons Live*, KFIV, Modesto/Stockton; Sue Stein, features editor, *Fate Magazine*; Michael Dukes, *Mornings with Michael Dukes*, KBYR, Anchorage; Gary Wellings, editor, *The Courant*, Ann Arbor; John Gomez, *John Gomez Show*, LI News Radio, New York; Rick Wiles, *TruNews*; Vinnie Penn, *Vinnie Penn Project*, New Haven; Rachel Alexander, editor, *Intellectual Conservative*; United Teacher, Los Angeles; Jim Robertson, editor, *Columbia Daily Tribune*, Columbia, MO; Scott Welton, reporter, *Standard-Democrat*, Cape Girardeau, MO; Dick Hughes, *Statesman-Journal*, Salem; Chuck "Nasty Man" Naste, KLSX, Los Angeles; KPSI, Los Angeles; KHJ, Los Angeles; Suzy Buchannan, editor, *Anchorage Press*; David Fox, book reviewer, *Anchorage Press*; Tara Servatious, host, *Tara Servatius Show*, WTMA, Charleston; Bill Meyers, *Bill Meyers Show*, KMED, Medford, OR; Bill Buckmaster, *Buckmaster Show*, KVOI, Tucson; Faune Riggin, *Morning News*, KZIM & KSIM, Cape Girardeau, MO; Gary Mantz and Suzanne Mitchell, *Gary Mantz Show*, KKNW, Seattle; Allan Handelman, *Allan Handelman Show*, WZTK, Charlotte; David Blomquist, *Bloomdaddy Show*, WJAS, Pittsburgh and WWVA, Wheeling; Drew Lane and Marc Fellhauer, WMGC, Detroit; Louie Free, host, *Brainfood for the Heartland*, Youngstown; Pat Thurston, *Pat Thurston Show*, KGO, San Francisco; KFKA, Newstalk, Denver/Ft. Collins; Ted Adams, Richard Dugan and Mike Williams, AM 1290, Santa Barbara; *Ideas & Discovery Magazine*; Patrick Huyghe, editor, *Edge Science Magazine*; Theresa Moreau and Jim Laris, *Pasadena Weekly*; Nanette Cazier, University of the Pacific; Denise Poon, executive producer, KCBS-TV, Los Angeles; Brenda Pavillian, news director, KDKA-TV, Pittsburgh; Don Miller, editor, *Monterey Herald*; Carolyn Gilbert, *Speak Out America, LA 36*; Lawrence McConnell, executive editor, *Roanoke Times*; Patrick Huyghe, editor, *The Anomalist*; Dr. John Mosley, program director, Griffith Observatory, Los Angeles; Tania Soussan, reporter, *Pasadena Star-News*; Jon Olson, chairman, Department of Anthropology, California State University, Los Angeles; Professor and Dr. Cristiano Fidani, University of Perugia, Andrea Bina Seismic Observatory, Italy; Paula Starr, assistant executive director, Southern California Indian Center; Patrick So, astronomical lecturer, Griffith Observatory, Los Angeles; Andrew Lindberg, producer, *KDKA Morning News*, Pittsburgh; Shawn Taylor, executive producer, *Thom Hartmann Program*; Stephanie Menendez, executive producer, *Bill and Wendy Show*, WGN, Chicago; Marshall Washburn, guest coordinator, *TruNews*; Sheryll Lamothe, assistant bureau chief, *Inside Edition*; Tom Lucero, producer, KFKA, Denver/Ft. Collins; Alice Rowley, editor, *Pittsburgh Post-Gazette*; Les Encinoca, producer, *McIntyre in the Morning*, KABC, Los Angeles; Leadership Team, Coordination Centre of Disaster Medicine, Brno, Czech Republic; Michael Figliola, booking producer, MSNBC, *News Nation with Tamron Hall*; Portalnet, Santiago, Chile; Dori O'Neal, arts and entertainment editor/book reviewer, *Tri-City Herald*, Kennewick, WA; Sarah Rooney, community programs manager, Heinz History Center/Affiliate Smithsonian Institution, Pittsburgh; Paula Johnson, *Book Buzz, Pasadena Weekly*; Anne Strieber, producer, *Dreamland Radio Show*; Leadership Team, Bright Claim, Atlanta; Steven Le Blanc, ex-submariner, United States Navy, PhD candidate, mathematics; Lisa Lyon, producer, *Coast to Coast AM*; Dr. Eric Force, Department of Geosciences, Distinguished Professor of Paleolimnology, University of Arizona, Tucson; Dave Malarkey, show host, *It's*

Your Turn, WISR, Pittsburg; Kristyn Clarke, reporter, *PCM World News*; Ben Gebhardt, associate editor, *John Carroll University News*, Cleveland; Brian Feinblum, editor, *Book Marketing Buzz*; Dr. Paul Burton, Burton Digital Books; Charlene Haislip, my sister, for all her expert help.

It is hoped that all the names listed above are correctly referenced; it was a difficult task to reconstruct the past twenty years of media venues, and press coverage; any errors are regretted. Lastly, it's not possible to list the names of all the managers of the many dozens of Waldenbooks, B.Dalton, Barnes and Noble, Doubleday, Borders, and other independent book stores and book fairs who invited me to speak to their clientele and sign books. I'll only say that Thomas Jefferson's quote about preferring newspapers to a government, should one have to choose, applies equally to purveyors of books. I must however single out one venue for a special note of thanks. It was a great honor to speak to the troops of the United States Marine Corps' Twenty-Nine Palms Base, even if it *was* in the Mojave Desert in the middle of the summer. Not only had I never witnessed a more respectful and courteous audience at a book signing, but heard "sir" uttered in my direction sufficient to last a lifetime. To the officers and marines at Twenty-Nine Palms: *semper fidelis*.

Index

P

Q

R

S

About the Author

David Nabhan is the author of three prior books concerning seismic forecasting, along with dozens of newspaper and magazine op-ed commentaries regarding seismicity, hydraulic fracking, climate issues, and other important matters. He is also a science fiction writer: *The Pilots of Borealis* (2015), Skyhorse Publishing/Talos Press, New York. Many dozens of interviews with and articles about Mr. Nabhan and his books, his published opinion pieces, along with contact links and other information can be accessed at www.earthquakepredictors.com.